# Introductory photochemistry

*Consulting Editor*

**P. Sykes,** M.Sc., Ph.D.
*Fellow of Christ's College*
*University of Cambridge*

---

*Other titles in the European Chemistry Series*

**Banwell:** Fundamentals of Molecular Spectroscopy
**Barnard:** Theoretical Basis of Inorganic Chemistry
**Barnard & Chayen:** Modern Methods of Chemical Analysis
**Barnard & Mansell:** Fundamentals of Physical Chemistry
**Benson:** Mechanisms of Inorganic Reactions in Solution
**Brennan & Tipper:** A Laboratory Manual of Experiments in Physical Chemistry
**Bu'lock:** The Biosynthesis of Natural Products
**Chapman:** Introduction to Lipids
**Garratt:** Aromaticity
**Gimblett:** Introduction to the Kinetics of Chemical Chain Reactions
**Hallas:** Organic Stereochemistry
**Kemp:** Practical Organic Chemistry
**Kemp:** Qualitative Organic Analysis
**Krebs:** Fundamentals of Inorganic Crystal Chemistry
**Sutton:** Electronic Spectra of Transition Metal Complexes
**Williams:** Introduction to the Chemistry of Enzyme Action
**Williams & Fleming:** Spectroscopic Methods in Organic Chemistry
**Williams & Fleming:** Spectroscopic Problems in Organic Chemistry

# Introductory photochemistry

**A. Cox**
*Lecturer, School of Molecular Sciences,*
*University of Warwick, Coventry*

**T. J. Kemp**
*Senior Lecturer, School of Molecular Sciences,*
*University of Warwick, Coventry*

McGRAW-HILL · LONDON

New York · St Louis · San Francisco · Düsseldorf
Johannesburg · Kuala Lumpur · London · Mexico
Montreal · New Delhi · Panama · Rio de Janeiro
Singapore · Sydney · Toronto

19,120

*Published by*

McGRAW-HILL Book Company (UK) Limited

MAIDENHEAD · BERKSHIRE · ENGLAND

07 094168 8



PRINTED AND BOUND IN GREAT BRITAIN

# Preface

Since Bowen's *Chemistry of Light* published in 1946, there has been no attempt to present a similar broad picture of the chemical effects of light for the purpose of undergraduate teaching, although the rapid growth of the subject in all its facets has provided a need for a text both introductory and up-to-date. There has been no lack of research monographs on specialized aspects of the subject, and also there has appeared the uniquely comprehensive *Photochemistry* of Calvert and Pitts. Our aim has not been to reiterate these achievements but instead, while drawing the attention of the student to the range of the subject, to emphasize those features which have been particularly significant for its further development and the inter-relation between the application of newer physical methods, the development of theoretical treatment of reaction mechanisms and intermediates, and their implications for synthesis. In chapters 1 and 2 are laid the foundations for discussion of the subject, namely the formulation of energy levels for molecules and the initial population of these following light absorption. Chapters 3 and 4 cover the reactions, both chemical and physical, of small molecules, inorganic ions and complexes, and simple and complex organic molecules. Chapter 5 covers the experimental aspects of the subject whilst the final chapter deals with topics which are directly related to photochemistry, e.g., vision and the photographic process. SI units have been employed as have the conventions → for emission and ← for absorption processes.

Our task has been eased to an extent by our having drawn on lectures delivered to final year honours students and by our having worked in the research laboratories of Professors F. S. Dainton, FRS, and D. H. R. Barton, FRS, to whom we acknowledge our very considerable debts.

A special note of thanks is paid to Dr J. A. Barltrop for undertaking the onerous task of reading the entire manuscript and offering invaluable criticism. Drs H. D. Burrows and D. M. Hirst of this department are also thanked for reading sections of the book.

School of Molecular Sciences
University of Warwick
September 1970

A. Cox
T. J. Kemp

# Contents

# 1 Introduction

## 1.1 Historical

Photochemistry, as the name suggests, is the study of chemical changes initiated by light, normally of visible or ultra-violet character but on occasion to the extremities of the range 10–2000 nm. The subject is related, therefore, to a number of other branches of chemistry, particularly electronic spectroscopy, which deals with the determination and quantum mechanical description of the levels involved in photochemical excitation, and also radiation chemistry, which is concerned with the chemical effects of radiation of $\lambda < 10$ nm (a rather arbitrary borderline, as will become apparent). Also closely involved with photochemistry are infra-red spectroscopy, which provides information on the band of vibrational states between the various electronic states, reaction kinetics, which furnishes a basis for discussion of the complex reaction systems often built up during photoirradiation, and electron spin resonance spectroscopy, which enables unequivocal identification of the ubiquitous paramagnetic intermediates following photodissociation. The photochemistry of macromolecules, particularly in assemblies, is the basis for any discussion of the photosynthetic process which is the chemical channel through which solar energy is adapted for the needs of higher forms of life.

Historically several forms of photochemical action have been noted, including the susceptibility of dyestuffs to fading in strong sunlight and the darkening of silver halide crystals. The light-sensitivity of a substantial number of simple inorganic and organic compounds had been commented upon before 1900 and some light-induced chemical transformations described, for example, the photodimerizations of anthracene in solution[1] and of solid α-*trans*-cinnamic acid to give dianthracene and (chiefly) α-truxillic acid respectively[2,3] (sections 4.3.4 and 4.3.5). The explosive combination of mixtures of hydrogen and chlorine induced by bright light was investigated by Draper[4] in 1844 and several of the main kinetic features had been determined by the turn of the century.[5] The science of photography was becoming established in several countries in the 1830's.[6]

Formulation of photochemical theory began in the first half of the nineteenth century with the Grotthus–Draper law, i.e.,

*only light* absorbed *by a molecule can induce a photochemical change within it.*

(Processes dependent upon transfer of energy or charge from a photoexcited molecule to a ground-state molecule, which then reacts, are excluded from this law.) Following the development of quantum theory, a more exact relation between absorption and reaction appeared in the Stark–Einstein (or Einstein) law:

*A molecule undergoing photochemical change does so through the absorption of a single quantum of light.*

The latter statement refers to the primary process of any reaction; species produced in this process may induce chain reactions involving the consumption of many further molecules of the starting material without any further light absorption. An exception is provided by so-called 'biphotonic processes' in which certain molecules take up *two* quanta successively to undergo particularly endothermic reactions.

The Stark–Einstein law leads directly to the index of efficiency of photochemical reaction, namely the *quantum yield*, denoted $\varphi$. This may be defined as:

$$\varphi = \frac{\text{number of molecules undergoing the particular process concerned}}{\text{number of quanta absorbed}}$$

If, as is more usually the case, several competing processes, 1, 2, . . . , $i$ operate following uptake of light by a molecule, then $\Sigma_i \varphi_i = 1$.

## 1.2 Energy levels

A molecule which has absorbed a quantum of light of wavelength 253·7 nm has received the equivalent of 469 kJ of energy, which is of the order of many single-bond energies, with the consequence that simple bond dissociation is one possible fate. Many molecules do not, however, dissociate on excitation, but undergo one or more of many further reactions and any attempts to correlate behaviour with structure must begin with a analysis of the nature of the molecular orbitals of a given molecule and an assessment of the likelihood of populating these orbitals on photoexcitation.

Calculation of the energy levels of an atom or molecule is performed by substituting into the Schrödinger wave equation a value for the potential energy of the assembly of nuclei and electrons concerned. Exact solution is possible only in the simplest cases, e.g., the hydrogen atom, and recourse is made to trial solutions based initially on an idealized and therefore approximate potential energy description for the system. The success of a particular trial solution is gauged by the degree to which the total energy of the system calculated thereby compares with those obtained by other trial solutions, the basis for this procedure being the variation theorem which states that:

*the value of the total energy* E *obtained from an approximate wave function is more positive, i.e., the system appears less stable, than the value obtained from the true wave function. That trial wave function giving the lowest value of* E *is adjudged the 'best.'*

The results with atoms are of great interest because the potential energy term is much simpler, embracing a single central positive charge and spherically symmetric wave functions for the filled shells of electrons. The exact solution of the Schrödinger equation for the hydrogen atom is particularly illustrative, indicating the necessity of introducing three quantum number into the various correct solutions or eigenfunctions;

$n$ *The principal quantum number.* This is the principal factor determining the energy of the orbital (especially so in the case of hydrogen-like atoms) and is correspondingly related to the dimension of the orbital, by which is meant that part of the orbital containing 0·9 of the electron's distribution. An orbital of quantum number $n$ shows $n - 1$ spherical nodes. $n$ can have any integral value, $1, 2, \ldots, \infty$.

$l$ *The azimuthal or orbital quantum number.* This governs the orbital angular momentum of the electron, a vector quantity, the magnitude of which is given by $[l(l+1)]^{\frac{1}{2}}\hbar$ ($\hbar = h/2\pi$). It also determines the shape of the orbital and an orbital of quantum number $l$ shows $l$ nodal planes. For a given value of $n$ there are corresponding $l$-values,

$$n-1, n-2, \ldots, 1, 0$$

For atoms other than hydrogen the set of $l$-values for orbitals of given $n$-quantum number are non-degenerate, the energy *increasing* with increasing $l$-value.

$m$ *The magnetic quantum number.* In the absence of an externally applied magnetic field each orbital of given values of $n$ and $l$ is $(2l+1)$-fold degenerate. $m$ can assume values

$$l, l-1, l-2, \ldots, 0, \ldots, -(l-1), -l$$

This degeneracy is removed on application of a magnetic field with a consequent $(2l+1)$-fold splitting of the level, the so-called Zeeman effect.

A further quantum number $s$ was introduced to account for fine structure of atomic spectra and for the magnetic properties of beams of atoms with zero orbital angular momentum. $s$ is designated the spin quantum number of the electron which is associated with a spin angular momentum vector of magnitude $[s(s+1)]^{\frac{1}{2}}\hbar$ and a magnetic moment $2[s(s+1)]^{\frac{1}{2}}$ Bohr magnetons. $s$ has a value $\pm 1/2$ and its existence has been justified subsequently by the introduction of relativistic wave mechanics.

The predicted spectroscopy of the hydrogen atom is exceptionally simple because all the $l$-states associated with a given value of $n$ are degenerate. However, the exact forms of $l$-functions give us an index of the probability of transition occurring between known levels because for any atom this is proportional to

$$\left(\int \psi_i \boldsymbol{\mu} \psi_f \, d\tau\right)^2 = \mathbf{m}^2 \tag{1.1}$$

where $\psi_i$, $\psi_f$ are the total wave functions for the initial and final states of the system and $\boldsymbol{\mu}$ is the dipole moment *operator* defined by

$$\boldsymbol{\mu} = \sum_{i=1}^{n} e r_i$$

where $e$ is the electronic charge and $r_i$ is the distance of the $i$th electron from the nucleus. $\mathbf{m}$ is called the transition moment integral and measures the displacement of charge taking place during the transition. Normally only a single electron is regarded as involved in an transition in either an atom or a molecule and the remaining electrons make no contribution to $\mathbf{m}$. Accordingly, $\psi_i$ and $\psi_f$ can be replaced by single-electron wave functions for the initial and final states for the electron undergoing promotion. It transpires that transitions of the type $l = 0 \rightarrow l = 0$, $l = 1 \rightarrow l = 1$, etc., have $\mathbf{m} = 0$ and are designated 'forbidden'. Also forbidden are transitions involving any change in spin angular momentum. For hydrogen the allowed transitions ($\mathbf{m} \neq 0$) can be summarized in the form of selection rules

$$\Delta n = \pm \text{ (any integer)}, \quad \Delta l = \pm 1$$

and all absorptions from the ground 1$s$ orbital ($n = 1, l = 0$) must be to $np$ orbitals ($n \geqslant 2, l = 1$). Further absorptions take place from occupied 2$p$ states, etc.

For atoms other than hydrogen, exact solutions of the Schrödinger equation are not feasible but instead hydrogenic wavefunctions, modified by the inclusion of additional parameters such as a screening constant, are tried until the total energy is minimized. The general character of the wave functions is maintained, however, the angular part remaining identical along with the selection rules, except that for orbitals of a given $n$-value the degeneracy of the $l$-states is removed such that the sub-state $l = 0$ (an $s$ state) is of lower energy than that of $l = 1$ (a $p$ state) which is, in turn, lower than the $l = 2$ state (a $d$ state), i.e., in general:

$$\text{energy:} \quad ns < np < nd < nf \ldots$$

This effect, due to greater 'penetration' of sub-shells by $s$ than $p$ wavefunctions, etc., gives rise to the approximately equal energies of 3$d$ and 4$s$ levels recognized in spectroscopy of transition metal ions. The spectroscopy of polyelectronic atoms with a single electron in an unfilled $l$ state is essentially that of a hydrogen atom with the modifications mentioned. For those with more than one electron in the unfilled $l$ state a new complication is introduced. Two separate electrons (1) and (2) with quantum numbers $(l_1, s_1)$ and $(l_2, s_2)$ respectively interact and their orbital and spin angular momenta couple to produce a resultant total angular momentum. For the present example this coupling can come about in two ways, namely,

(a) *L–S* coupling; the orbital components can couple together to produce a net total orbital angular momentum characterized by a quantum number $L$, and the spin components can couple together to produce a net total spin angular momentum characterized by a quantum number $S$. $L$ and $S$ then couple together to produce an angular momentum of quantum number $J$. It is important to realize the number of ways available of vectorially adding spin and angular momenta. The orbital angular momentum of electron (1) can lie anywhere between

$$l_1, l_1 - 1, \ldots, 0, \ldots, -(l_1 - 1), -l_1$$

corresponding to all the possible values of $m_{l_1}$; again $s_1$ can be $-1/2$ or $+1/2$. Similar considerations apply to electron (2) and furthermore all ways of combining $L$ and $S$ must also be considered. This type of coupling, which also known as Russell–Saunders coupling, is important for light elements, and some examples follow.

(b) *j–j* coupling. In this case each momentum $l_i$ couples with the corresponding spin momentum $s_i$ to give a resultant angular momentum $j_i$; these resultant momenta are then further coupled to give the resultant $J$.

It appears from the foregoing discussion that the various modes of coupling imply the existence of many electronic states for an atom containing only two electrons in unfilled shells, and this is apparent in the multiplet structure of spectral lines. However, certain trends are apparent which simplify the situation and these are best illustrated by example.

### *Example 1.1*

Two inequivalent electrons; $l_1 = 1, l_2 = 2$. These can be combined vectorially to give $L$ values of $(2 + 1)$, $(2 - 1)$ and all integral values between these, i.e., to give values of 3, 2, 1. The angular momenta associated with these are given by $L^{\frac{1}{2}}(L + 1)^{\frac{1}{2}}\hbar$, i.e., $12\,\hbar$, $6\,\hbar$, and $2\,\hbar$. For two electrons there will always be $2\,l_{min} + 1$ values of $L$ where $l_{min}$ is the smaller $l$-quantum number.

Considering now the spin quantum numbers $s_1$ and $s_2$; the only possibilities here are

$$S = s_1 + s_2, \quad s_1 + s_2 - 1$$

i.e.,

$$S = 1 \quad \text{or} \quad 0$$

The $S = 1$ and $S = 0$ states are denoted triplet and singlet states respectively.

Coupling of the momenta corresponding to the values of $L$ and $S$ is performed either by drawing vector diagrams or by writing down all values differing by unity between $(L + S)$ and $(L - S)$. In the present case these are

(a) $L = 3; J = 3 + 1, 3, 3 - 1$ (for $S = 1$) and 3 (for $S = 0$)
(b) $L = 2; J = 2 + 1, 2, 2 - 1$ (for $S = 1$) and 2 (for $S = 0$)
(c) $L = 1; J = 1 + 1, 1, 1 - 1$ (for $S = 1$) and 1 (for $S = 0$)

Using the spectroscopic convention of expressing states in the form

$$^{2S+1}L_J$$

then the two electrons give rise to the states;

$$^3F_4\ ^3F_3\ ^3F_2\ ^3D_3\ ^3D_2\ ^3D_1\ ^3P_2\ ^3P_1\ ^3P_0 \quad \text{and} \quad ^1F_3\ ^1D_2\ ^1P_1$$

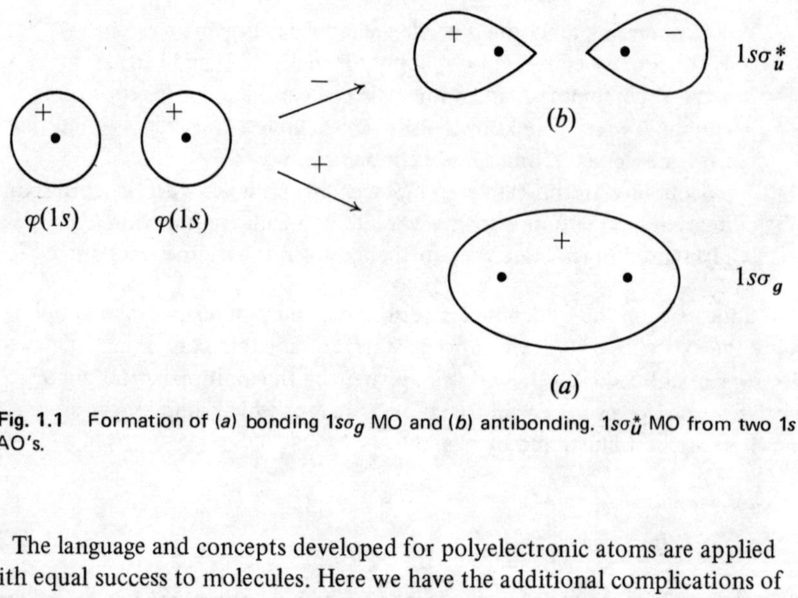


**Fig. 1.1** Formation of (*a*) bonding $1s\sigma_g$ MO and (*b*) antibonding. $1s\sigma_u^*$ MO from two 1*s* AO's.

The language and concepts developed for polyelectronic atoms are applied with equal success to molecules. Here we have the additional complications of quantization of rotational and vibrational energy as well as electronic energy, but fortunately the acceptance of the Born-Oppenheimer approximation, i.e., that the great differences between $E_{\text{elec}}$, $E_{\text{vib}}$, and $E_{\text{rot}}$ implies that one can treat them quite separately, vastly simplifies the problem. The quantization of rotational and vibrational energies are dealt with in every elementary spectroscopy book and are of concern to the photochemist inasmuch as vibrational spectra or vibrational structure on electronic spectra give information on the shapes of the potential energy curves of molecules in ground and excited electronic states.

Electronic states of molecules are best discussed in terms of molecular orbital (MO) theory, of which a few salient features are now given. MO's are normally constructed from AO's of the component atoms of the molecule by taking appropriate linear combinations of those of the latter which have compatible symmetries (the LCAO approximation). For the $H_2$ molecule the lowest lying MO's are

$$\psi_+ = \varphi(1s) + \varphi(1s)$$
$$\psi_- = \varphi(1s) - \varphi(1s)$$

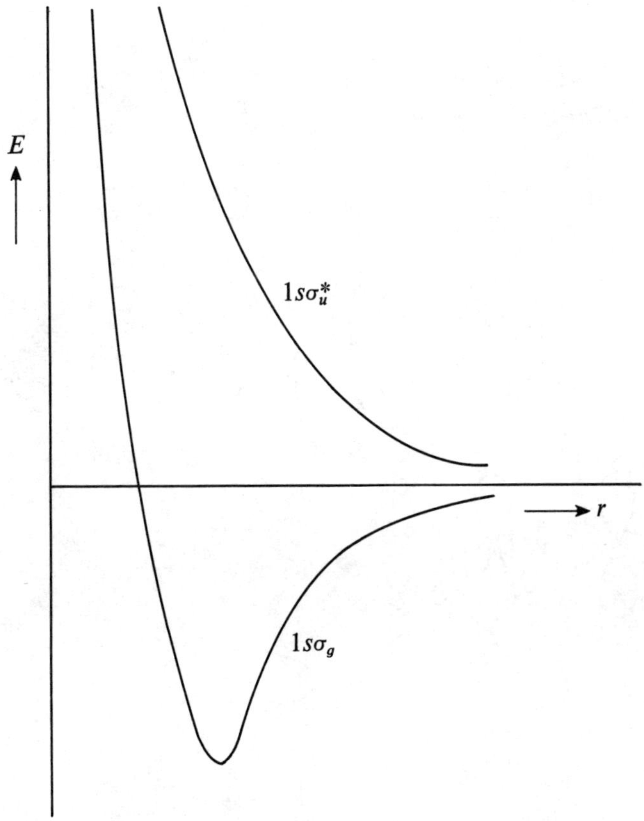


**Fig. 1.2** Morse curves for $1s\sigma_g$ and $1s\sigma_u^*$ MO's.

where the AO's are normalized, i.e., $\int\psi^2\,d\tau = 1$. $\psi_+$ corresponds to a summing of AO's of positive sign, i.e., to a build up of charge density between the nuclei to give a bonding MO denoted $1s\sigma$ (Fig. 1.1 (*a*)), being the ground state of the molecule. (Details of molecular quantum numbers are given below.) $\psi_-$ corresponds to a cancellation of charge density between the nuclei resulting in enhanced nuclear repulsion and an antibonding MO denoted $1s\sigma^*$ (Fig. 1.1(*b*)). The Morse curves for these MO's are illustrated in Fig. 1.2, demonstrating that excitation to the $1s\sigma^*$ orbital from the ground state leads to dissociation to two ground-state hydrogen atoms with a large kinetic energy. Various combinations of other AO's of hydrogen lead to additional MO's exemplified in Fig. 1.3.

The classification of molecular states of diatomic molecules is closely analogou

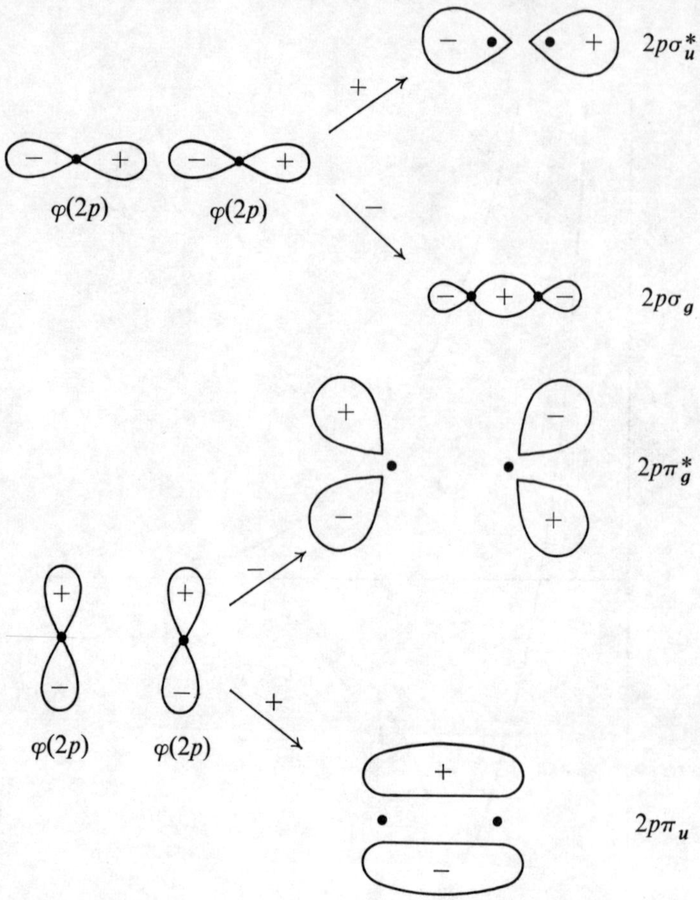


**Fig. 1.3** Formation of $2p\sigma$ and $2p\pi$ MO's from $2p$ AO's.

to that of atoms except that in the internuclear axis we have a natural direction of reference for considering the quantization of orbital angular momentum. The axial component of the latter is quantized, with quantum numbers $\lambda = 0, 1, 2, 3, \ldots$, the states corresponding to these values being symbolized $\sigma, \pi, \delta, \varphi, \ldots$ respectively. For polyelectronic systems we are concerned, as with atoms, with the *total* component of the orbital angular momenta, denoted $\Lambda$, which is arrived at by summing individual $\lambda$ values, taken as positive or negative, in all possible ways to give a positive value for $\Lambda$, e.g., for a $\pi$ and a $\delta$ electron ($\lambda_1 = 1$, $\lambda_2 = 2$) we obtain $\Lambda = 3$ or 1. States for which $\Lambda = 0, 1, 2, 3, \ldots$ respectively are called $\Sigma$, $\Pi$, $\Delta$, and $\Phi$ respectively.

Spin angular momentum is not greatly affected by the internuclear field and exactly the same terminology is used for molecules as for atoms. Summation of spin and orbital angular momentum, the analogue of Russell–Saunders coupling in atoms, is achieved by treating the axial components of both momenta in the usual way to give a resultant denoted $\Omega$, the analogue of $J$. Accordingly the molecular term symbol becomes

$$^{2S+1}\Lambda_\Omega$$

Two further data incorporated into the term symbol concern the symmetry of the MO. If the diatomic is homonuclear then the operation of inversion ($i$) through the centre of symmetry of the MO either leaves the sign of the wave function unchanged (as in the $1s\sigma$ MO of $H_2$), in which case the MO is described as *even* or $g$, or it reverses the sign (as in the $1s\sigma^*$ MO) and the MO is labelled *odd* or $u$. In any diatomic the operation of reflection ($\sigma_h$) through the plane of symmetry again leaves the sign of the wave function unchanged (symmetrical or +) or reversed (antisymmetrical or –). Accordingly the ground state of $H_2$ can be written $(1s\sigma)_g{}^2\ {}^1\Sigma_g^+$.

Other singlet states of hydrogen, i.e., with $2S + 1 = 1$, are obtained by considering excitation of one electron to ($a$) a $2s\sigma_g$ MO to give a $(1s\sigma_g 2s\sigma_g)\ {}^1\Sigma_g^+$ state, ($b$) a $2p\sigma_g$ MO to give a $(1s\sigma_g 2p\sigma_g)\ {}^1\Sigma_u^+$ state and ($c$) a $2p\pi_u$ MO to give a $(1s\sigma_g 2p\pi_u)\ {}^1\Pi_u$ state. The energy sequence of these can be determined qualitatively by considering the degree of overlap of the MO's and by applying Hund's rule. Orbitals formed from $2p$ and $1s$ electrons give better overlap than those from $1s$ and $2s$ electrons, and we have therefore the sequence

$$^1\Sigma_u^+ < {}^1\Pi_u < {}^1\Sigma_g^+$$

Selection rules for transitions between these states are obtained by applying equation (1.1);

(a) $\Delta\Lambda = 0, \pm 1$

(b) $\Delta S = 0$

(c) Symmetry restrictions:

$$\Sigma^+ \leftrightarrow \Sigma^+, \Sigma^- \not\leftrightarrow \Sigma^-, \Sigma^+ \not\leftrightarrow \Sigma^-$$

$$g \leftrightarrow u,\ g \not\leftrightarrow g,\ u \not\leftrightarrow u$$

The designations $\sigma$ and $\pi$ in non-linear polyatomic molecules do not refer to the symmetry of the whole molecule but to local symmetry; the electron density for a $\sigma$ orbital is greatest along the bond axis whereas for a $\pi$ orbital the bond axis lies in a nodal plane of zero electron density. The general order of orbital energies is shown in Fig. 1.4 from which it is apparent that $\pi^* \leftarrow n$ transitions lie at longest wavelength with $\pi^* \leftarrow \pi$ at intermediate, and $\sigma^* \leftarrow \sigma$ transitions at shortest, wavelengths. Relatively few types of molecules have only $\sigma^* \leftarrow \sigma$

transitions of which the alkanes are the most conspicuous example. With small polyatomic molecules one is concerned largely with the geometry of the ground and excited states for the state of hybridization of the central atom in these states will determine their energy. The spectroscopy and photochemistry of larger molecules are determined by the degree of interaction between electrons at different points in the molecular framework. It is found spectroscopically that electrons in $\pi$ bonds or in non-bonding orbitals do not interact when separated by more than one $\sigma$ bond but that strong interaction occurs when separation is confined to this limit. This is because of the shapes of $p$-orbitals which permit

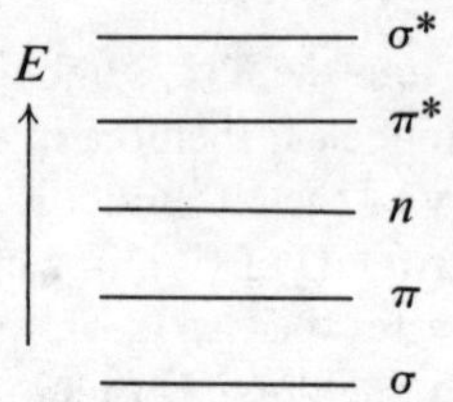


**Fig. 1.4** General order of orbital energies in non-conjugated systems (energy separations are unequal).

good overlap between adjacent pairs of a series of carbon atoms in this type of system, e.g., in polyenes, such as butadiene, resulting in the formation of a set of

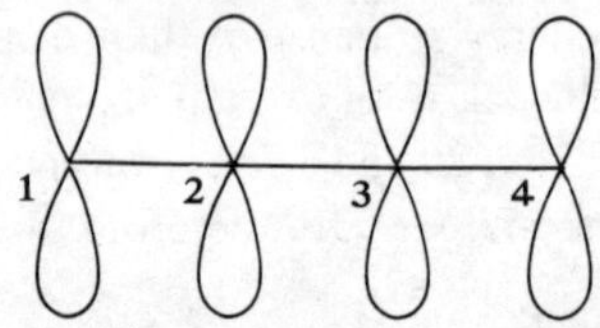


MO's, one of which involves bonding between $C_2$ and $C_3$ and is occupied. Theoretical treatments of such a system depend on the assumption that $\sigma$ orbitals are non-interacting either with each other or with the $\pi$ orbitals. The simplest approach postulates that the $\pi$ electrons are effectively in a one-dimensional box the length of the $\sigma$ framework ($L$) and of zero potential at all points but of infinite potential at the extremities, occupying the resulting orbitals in pairs. An exact solution for this problem can be obtained using the Schrödinger equation producing energy levels $\epsilon_n = n^2h^2/8\,mL^2$ where $n$ is the quantum number (1, 2, 3, . . . ), $h$ is Planck's constant and $m$ is the electronic mass. This method, called the free electron (FE) theory, accounts successfully for the absorption spectra of a number of long-chain conjugated molecules but is is far less widely applied than the LCAO approach for other types of molecule. In the latter, MO's for a series of carbon $2p_z$ orbitals are made up as follows

$$\psi_j = c_{j1}\,\varphi_1 + c_{j2}\,\varphi_2 + \ldots + c_{jr}\,\varphi_r \ldots + c_{jn}\,\varphi_n \tag{1.2}$$

where $\psi_j$ is the $j$th MO, $\varphi_r$ is the AO for the $r$th atom and $c_{jr}$ is the coefficient of the $r$th AO in the $j$th MO. The problem is to obtain the coefficients and this is accomplished by utilizing the Variation Theorem (p. 3) in its explicit form

$$E = \frac{\int \psi \mathbf{H} \psi \, d\tau}{\int \psi^2 \, d\tau} > E_{\text{true}} \qquad (1.3)$$

$E$ is the energy obtained from the wave function $\psi$ by applying the Hamiltonian operator $\mathbf{H}$ in equation (1.3). The procedure is to minimize $E$, after substituting (1.2) into (1.3), by differentiating with respect to each of the coefficients $c_r$ in turn to yield a set of $n$ equations. In these appear the general integrals $\int \varphi_r \mathbf{H} \varphi_s \, d\tau$ and $\int \varphi_r \varphi_s \, d\tau$ which are shortened to $H_{rs}$ and $S_{rs}$ respectively. The Hückel approximation consists of taking all the terms $H_{rr}$, known as Coulomb integrals and representing the energy in a carbon $2p_z$ orbital, as equal (and denoted $\alpha$) and all the terms $H_{rs}$ $(r \neq s)$, known as resonance integrals and representing the energy of interaction of two AO's, as zero for non-adjacent atoms and equal for all pairs of adjacent atoms (and denoted $\beta$). Also $S_{rr} = 1$ and $S_{rs} = 0$ $(r \neq s)$. This simplifies the set of $n$ equations which have a non-trivial solution only if the corresponding secular determinant vanishes. Expansion of this determinant yields a polynomial equation with $n$ real roots of the form

$$(\alpha - E) = -m_j \beta \qquad j = 1, \ldots, n$$

or

$$E_j = \alpha + m_j \beta$$

This implies a set of $n$ energy levels spaced above and below an energy zero taken as $\alpha$. Since $\beta$ is negative, positive values of $m_j$ represent more negative energy levels than that of an electron in a single $2p_z$ orbital, i.e., bonding MO's, and $m_j = 0$ corresponds to a non-bonding MO. Negative values of $m_j$ lead to the higher energy (antibonding) levels. This situation is readily apparent from the relatively facile treatment of simple systems as follows.

### *Example 1.2*

HMO's for ethylene. The secular determinant is

$$\begin{vmatrix} H_{11} - S_{11} E & H_{12} - S_{12} E \\ H_{21} - S_{21} E & H_{22} - S_{22} E \end{vmatrix} = 0$$

and putting $H_{11} = H_{22} = \alpha, H_{12} = H_{21} = \beta, S_{11} = S_{22} = 1, S_{12} = S_{21} = 0$

then

$$\begin{vmatrix} \alpha - E & \beta \\ \beta & \alpha - E \end{vmatrix} = 0$$

and dividing each term by $\beta$

$$\begin{vmatrix} (\alpha - E)/\beta & 1 \\ 1 & (\alpha - E)/\beta \end{vmatrix} = 0$$

Putting $x = (\alpha - E)/\beta$, the determinant can be expanded to give $x^2 - 1 = 0$ and $x = \pm 1$ or $E_1 = \alpha + \beta$ and $E_2 = \alpha - \beta$ (Fig. 1.5).

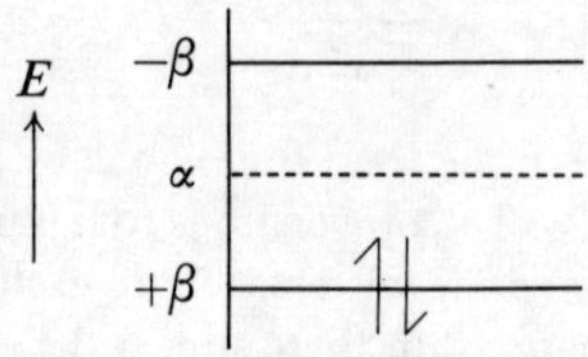


**Fig. 1.5** HMO's for $C_2H_4$.

*Example 1.3*

HMO's for the allyl system. With a wavefunction $\psi = c_1\varphi_1 + c_2\varphi_2 + c_3\varphi_3$ we obtain now a secular determinant after making the usual approximations, e.g., $H_{13} = H_{31} = 0$

$$\begin{vmatrix} \alpha - E & \beta & 0 \\ \beta & \alpha - E & \beta \\ 0 & \beta & \alpha - E \end{vmatrix} = 0$$

Putting $(\alpha - E)/\beta = x$ then the polynomial equation becomes $x^3 - 2x = 0$ which has roots $x = -\sqrt{2}, x = 0, x = +\sqrt{2}$, i.e., corresponding to

$$E_1 = \alpha + \sqrt{2}\beta$$
$$E_2 = \alpha$$
$$E_3 = \alpha - \sqrt{2}\beta$$

These levels are illustrated for various allyl species in Fig. 1.6.

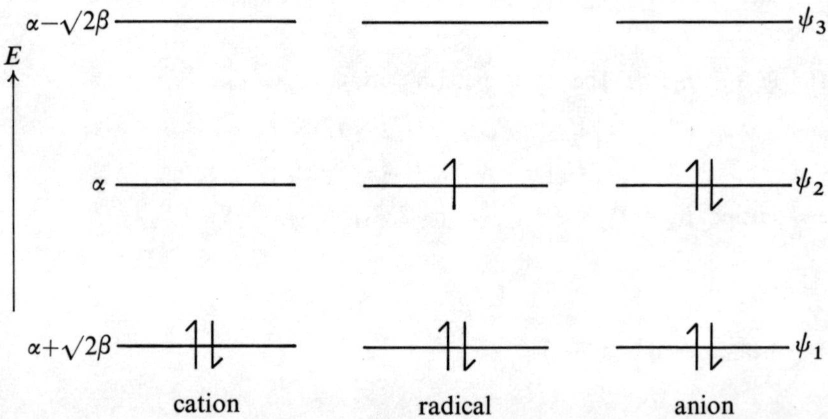


**Fig. 1.6** HMO's for the allyl system.

While further discussion of HMO's cannot be given here, it is clear that extensions to butadiene and benzene can be made. In general, the combination of $n$ AO's gives $n$ MO's with $\psi_j$ (numbering from lowest to highest energy) displaying $j-1$ nodal points. The calculation of $c_r$ depends on substitution of $E$ back into the set of simultaneous equations, e.g., for ethylene the determinant becomes

$$\begin{vmatrix} c_1(H_{11}-S_{11}E) & c_2(H_{12}-S_{12}E) \\ c_1(H_{21}-S_{21}E) & c_2(H_{22}-S_{22}E) \end{vmatrix} = 0$$

i.e., in terms of $x$

$$\begin{vmatrix} c_1 x & c_2 \\ c_1 & c_2 x \end{vmatrix} = 0$$

which corresponds to the equations

$$c_1 x + c_2 = 0$$
$$c_1 + c_2 x = 0$$

for $x = -1$ (lowest level)

$$-c_1 + c_2 = 0 \quad \therefore c_1 = c_2$$

But the normalization condition is $\sum_n c_r^2 = 1$

$$\therefore c_1^2 + c_2^2 = 1$$
$$\therefore c_1 = c_2 = 1/2$$

The MO's for inorganic complexes are obtained by combining AO's for the valence electrons of the metal atom with AO's or MO's of the ligands, denoted $\varphi_M$ and $\varphi_L$ respectively. Successful combination to give strong bonding depends on matching the component orbitals both as regards their energy and symmetry:

$$\psi_+ = c_1\varphi_M + c_2\varphi_L$$
$$\psi_- = c_1\varphi_M - c_2\varphi_L \qquad (1.4)$$

For an octahedral complex

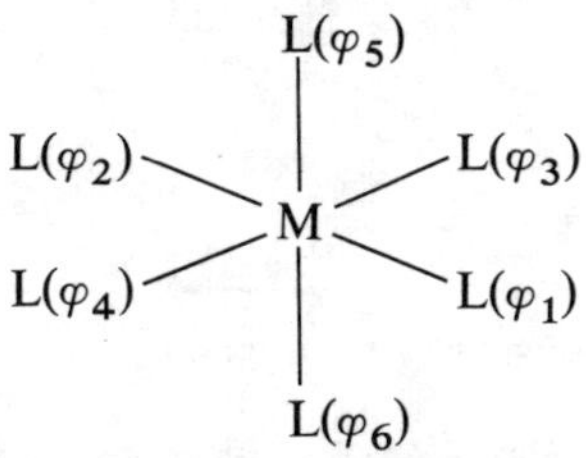


the matching of metal and ligand orbitals is given in Table 1.1 and energy levels are presented in Fig. 1.7 (the degeneracies are apparent from Table 1.1). The

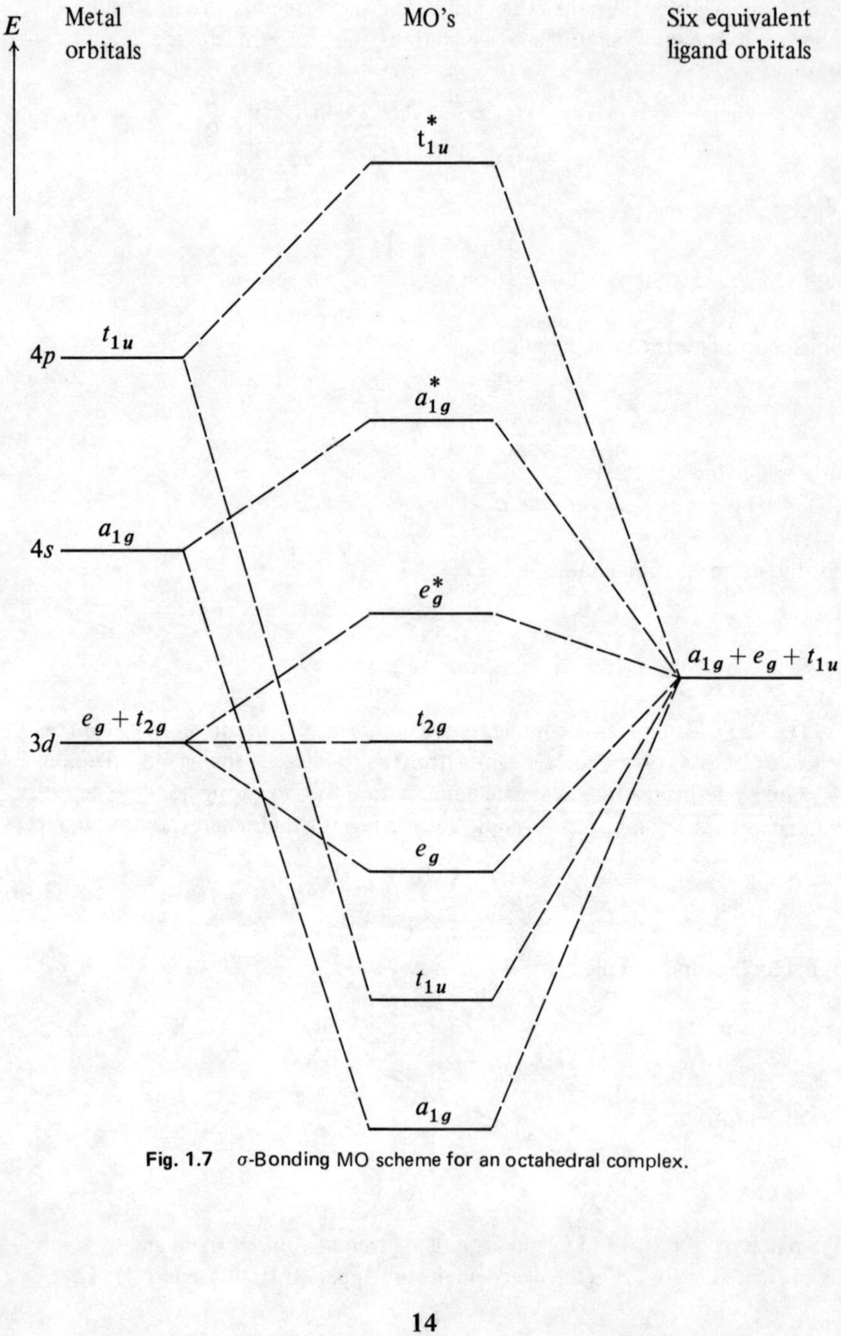


**Fig. 1.7** σ-Bonding MO scheme for an octahedral complex.

twelve electrons originating from the ligands are accommodated in the three lowest bonding MO's whilst the $t_{2g}$ orbitals remain essentially 'metal' in character. The low energy transitions of metal complexes occur between $t_{2g}$ and $e_g^*$ orbitals. $\pi$ orbitals of the ligand can easily be incorporated into an extended version of Fig. 1.7.

TABLE 1.1: Matching of metal and ligand MO's for $\sigma$-bonding in an octahedral complex

| $\varphi$M | Ligand combination | Species |
|---|---|---|
| $4s$ | $(1/\sqrt{6})(\varphi_1 + \varphi_2 + \varphi_3 + \varphi_4 + \varphi_5 + \varphi_6)$ | $a_{1g}$ |
| $4p_x$ | $(1/\sqrt{2})(\varphi_1 - \varphi_2)$ | |
| $4p_y$ | $(1/\sqrt{2})(\varphi_3 - \varphi_4)$ | $t_{1u}$ |
| $4p_z$ | $(1/\sqrt{2})(\varphi_5 - \varphi_6)$ | |
| $3d_{z^2}$ | $(1/\sqrt{12})(-\varphi_1 - \varphi_2 - \varphi_3 - \varphi_4 + 2\varphi_5 + 2\varphi_6)$ | $e_g$ |
| $3d_{x^2-y^2}$ | $(1/2)(\varphi_1 + \varphi_2 - \varphi_3 - \varphi_4)$ | |
| $3d_{xy}$ | — | |
| $3d_{yz}$ | — | $t_{2g}$ |
| $3d_{zx}$ | — | |

## 1.3 Group theory notation

It was noted above that combination of orbitals to give MO's can only be achieved when the symmetries as well as the energies are compatible. Accordingly it is convenient to classify electronic states in terms of their symmetry properties, using small letters for one-electron orbitals and large letters for many-electron states. The Mulliken terminology is based on the following rules:

(*a*) All non-degenerate species are designated $A$ or $B$; doubly degenerate species $E$ and triply degenerate species either $T$ or $F$.

(*b*) Non-degenerate species which are symmetric with respect to rotation $C_n^1$ about the principal $C_n$ axis are designated $A$: those species antisymmetric with respect to this rotation are designated $B$.

(*c*) Subscripts to $A$ and $B$ signify that the species is symmetric (subscript 1) or antisymmetric (subscript 2) with respect to a $C_2$ operation perpendicular to the principal axis, or, if the $C_2$ axis is absent, to a vertical plane of symmetry. Subscripts to $E$ and $T$ species are not easily derived, varying with the point group concerned.

(*d*) Primes and double primes mean that the species is symmetric or antisymmetric with respect to the operation $\sigma_h$.

(*e*) Subscripts $g$ and $u$ have their usual meaning, i.e., that the species is symmetric or antisymmetric with respect to the operation ($i$). The various symmetry symbols are defined in Table 1.2, and in Example 1.4 is outlined the use of a character table.

TABLE 1.2: Symmetry symbols

| Symbol | Symmetry element | Symmetry operation |
|---|---|---|
| $E$ | – | No change (identity) |
| $C_n$ | $n$-Fold axis of rotation (the principal axis of symmetry is that of largest $n$) | Rotation about an axis of symmetry by $360°/n$ |
| $\sigma_h$ | Plane of symmetry perpendicular to the principal axis of symmetry | Reflection in the plane of symmetry |
| $\sigma_v$ | Plane of symmetry containing the principal axis of symmetry | Reflection in the plane of symmetry |
| $\sigma_d$ | Plane of symmetry containing the principal axis of symmetry and bisecting the angle between two $a$-fold axes of symmetry which are perpendicular | Reflection in the plane of symmetry |
| $S_n$ | $C_n + \sigma_h$ | Rotation about an axis by $360°/n$ followed by a reflection in a plane perpendicular to the axis of rotation |
| $i$ | Centre of symmetry | Inversion in a centre of symmetry |

### *Example 1.4 Symmetry species for an octahedral complex*

Referring to the character table for the relevant point group, $O_h$ (Table 1.3), the point group and the relevant symmetry operations are given in the first row. The numbers preceding the latter refer to the number of operations of the same class giving rise to an identical vertical pattern of entries. The first column summarizes the effect of *all* the operations upon the wave function, for example, the $s$-wave function of a central metal atom is unchanged in sign after any operation and corresponds to the symbol $a_{1g}$. The remaining numbers, known as characters, refer to the result of the symmetry operation upon the wave function;

TABLE 1.3: Character table for point group $O_h$

| $O_h$ | $E$ | $8C_3$ | $6C_2$ | $6C_4$ | $3C_2$ | $i$ | $6S_4$ | $8S_6$ | $3\sigma_h$ | $6\sigma_d$ | | |
|---|---|---|---|---|---|---|---|---|---|---|---|---|
| $A_{1g}$ | 1 | 1 | 1 | 1 | 1 | 1 | 1 | 1 | 1 | 1 | | $x^2+y^2+z^2$ |
| $A_{2g}$ | 1 | 1 | −1 | −1 | 1 | 1 | −1 | 1 | 1 | −1 | | |
| $E_g$ | 2 | −1 | 0 | 0 | 2 | 2 | 0 | −1 | 2 | 0 | | $(2z^2-x^2-$ <br> $x^2-y^2)$ |
| $T_{1g}$ | 3 | 0 | −1 | 1 | −1 | 3 | 1 | 0 | −1 | −1 | $(R_x, R_y, R_z)$ | |
| $T_{2g}$ | 3 | 0 | 1 | −1 | −1 | 3 | −1 | 0 | −1 | 1 | | $(xy, yz, xz)$ |
| $A_{1u}$ | 1 | 1 | 1 | 1 | 1 | −1 | −1 | −1 | −1 | −1 | | |
| $A_{2u}$ | 1 | 1 | −1 | −1 | 1 | −1 | 1 | −1 | −1 | 1 | | |
| $E$ | 2 | −1 | 0 | 0 | 2 | −2 | 0 | 1 | −2 | 0 | | |
| $T_{1u}$ | 3 | 0 | −1 | 1 | −1 | −3 | −1 | 0 | 1 | 1 | $(x, y, z)$ | |
| $T_{2u}$ | 3 | 0 | 1 | −1 | −1 | −3 | 1 | 0 | 1 | −1 | | |

operations leaving $\psi$ unchanged and reversed in sign are designated 1 and $-1$ respectively. Degenerate wave functions are treated together in the form of a linear comb ination and the identity operation $E$, for example, produces characters 2 and 3 for doubly- and triply-degenerate functions. The remaining two areas of the table describe the symmetry species to which the Mulliken symbols refer. That next to the characters refers to the coordinates and rotations about the coordinate axes, whilst the other area shows various functions of the coordinates.

For the octahedral complex, the $s$-orbital is labelled $a_{1g}$ whilst the three $p$-orbitals, being triply degenerate and antisymmetric with respect to inversion, are denoted $t_{1u}$. The $d$-orbitals are symmetric towards inversion and are all labelled with the subscript $g$, the doubly-degenerate pair transforming as $e_g$ and the triply-degenerate orbitals as $t_{2g}$. Subscripts 1 and 2 relate to the operation $C_4$. In general the symmetry properties of a total wave function for a complex are closely similar to those of the one-electron orbital of the same orbital quantum number and $P$ and $D$ terms split in the same way as $p$- and $d$-orbitals.

## 1.4 Absorption and emission of light

In the preceding section the formulation of energy levels for different types of molecules has been discussed in terms of the appropriate eigenfunctions, and the selection rules for transitions between these levels have been given. These depend on substitution into equation (1.1) of the eigenfunctions concerned, although the result, i.e., whether the transition moment integral is zero (for a 'forbidden' transition) or non-zero (for an 'allowed' transition) can be arrived at without detailed calculation. The exact magnitude of **m**, however, involves calculation and the absorption characteristics of a molecule are normally discussed in more qualitative terms and are based on experimental determination.

The absorption of a molecule at a given frequency $\nu$ is related to its concentration $c$ (in mols $l^{-1}$) and the path length of the solution ($l$ cm) through the Beer–Lambert law

$$\log_{10}\frac{I_0}{I} = \epsilon c l \qquad (1.5)$$

where $I_0$ and $I$ are the intensities of the incident light beam at the face of the solution and at depth $l$ respectively and $\epsilon$ is denoted the decadic molar extinction coefficient, in units of $l\ mol^{-1}\ cm^{-1}$ and is characteristic of the molecule; the function $\log_{10}(I_0/I)$ is called the absorbance or the optical density of the solution. $\epsilon$ is a function of frequency, values quoted normally referring to the position of maximum absorption, and the intrinsic ability of a molecule to absorb is more adequately expressed by its oscillator strength $f$

which is related to the integral of $\epsilon$ over all frequencies of absorption by the equation

$$f = \frac{2303mc^2}{\pi Ne^2} \int_\nu \epsilon \, d\nu = 4{\cdot}319 \int_\nu \epsilon \, d\nu \tag{1.6}$$

where $e$ and $m$ refer to the electronic charge and mass respectively, $N$ is Avogadro's number and $c$ is now the speed of light; $\nu$ is expressed in $cm^{-1}$. It can further be shown that the transition moment **m** for an absorption is related to $f$ by the equation

$$f = \frac{8\pi^2 \, mc\nu\mathbf{m}^2}{3he^2} = 4{\cdot}704 \times 10^{29} \, \nu\mathbf{m}^2 \tag{1.7}$$

where $h$ is Planck's constant and **m** is in electrostatic units. If $f$ is unity for a strongly absorbing dye ($\nu_{max}$ 20 000 $cm^{-1}$ or 500 nm) then $m$ is $10^{-17}$ e.s.u. or 10 debyes.

A further relation exists between the absorption characteristics of an atom or molecule and its lifetime in the corresponding excited state, although slight elaboration of the following discussion is necessary for molecules. If the lower state (1) containing $n_1$ atoms is in equilibrium with an upper state (2) containing $n_2$ atoms in a field of exciting radiation of the appropriate frequency and of density $\rho$, then the number of absorption acts will be $n_1\rho B_{12}$ where $B_{12}$ is characteristic of the atom and is termed the Einstein absorption probability coefficient. Excited atoms in state (2) are presumed to radiate spontaneously in returning to state (1) with $n_2 A_{21}$ acts per second where $A_{21}$ is the Einstein emission probability coefficient. At equilibrium

$$n_1 \, \rho B_{12} = n_2 \, A_{21}$$

and

$$n_2/n_1 = e^{-h\nu/kT}$$

Inclusion of *stimulated* emission (the basis for laser action) from state (2) gives a further $n_2 B_{21} \, \rho$ acts of emission where $B_{21}$ is the appropriate probability coefficient, i.e., at equilibrium

$$n_1 \, \rho B_{12} = n_2(A_{21} + B_{21} \, \rho)$$

The Einstein coefficients are related to transition moments **m** as follows:

$$A_{21} = \frac{64\pi^4 \bar{\nu}^3 \mathbf{m}_{21}{}^2}{3hc^3} \tag{1.8}$$

$$B_{12} = B_{21} = \frac{8\pi^3}{3h^2} \mathbf{m}_{12}{}^2 \tag{1.9}$$

($\bar{\nu}$ is the true frequency of the transition in $s^{-1}$). Furthermore, from equation (1.7)

$$A_{21} = \frac{8\pi^2 e^2 \nu^2 f_{21}}{mc} \tag{1.10}$$

$$B_{12} = \frac{\pi e^2 f_{12}}{mhc^2 \nu} \tag{1.11}$$

where $\nu$ is in $cm^{-1}$. Now $A_{21}$ is the number of times per second that an atom in state (2) reverts to state (1) radiatively, i.e., with emission of radiation. The mean lifetime $\tau$ in the excited state is therefore

$$\tau = 1/A_{21} \tag{1.12}$$

and from equations (1.10) and (1.6) we find

$$\tau = \frac{me}{8\pi^2 e^2 \nu^2 f} = \frac{1{\cdot}50}{\nu^2 f} \tag{1.13}$$

and

$$\tau = \frac{N}{8\pi \times 2303\nu^2 c \int_\nu \epsilon \, d\nu} = \frac{3{\cdot}47 \times 10^8}{\nu^2 \int_\nu \epsilon \, d\nu}$$

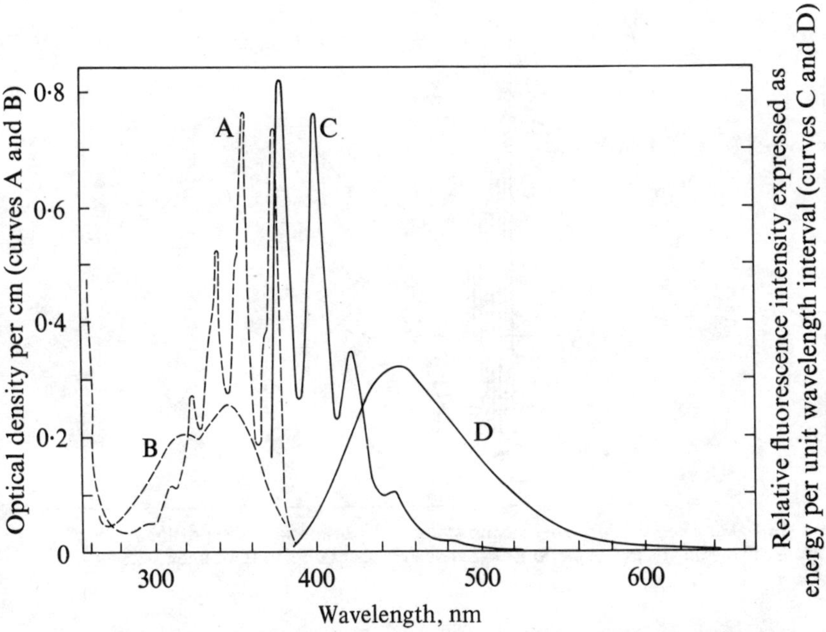


**Fig. 1.8** Absorption (broken lines) and fluorescence (full lines) spectra of solutions of anthracene (curves A and C) and quinine hydrogen sulphate (curves B and D) in ethanol and 0·05 M aqueous $H_2SO_4$ respectively.[7]

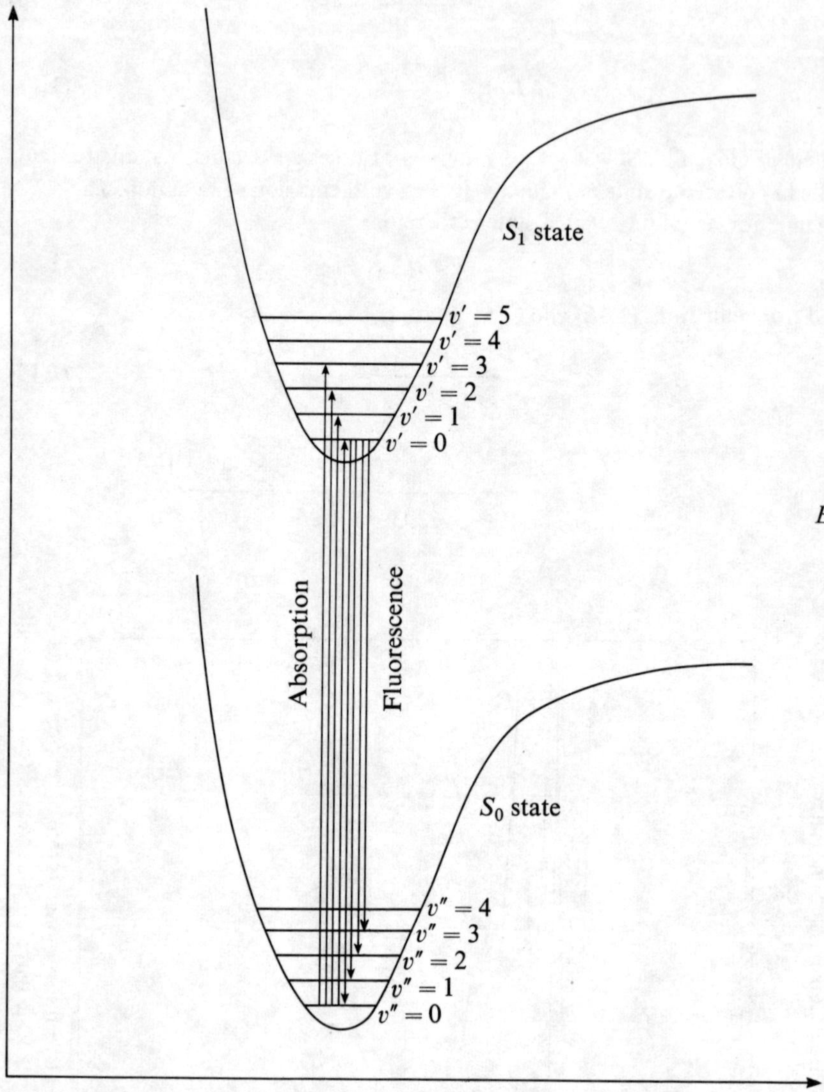


**Fig. 1.9** Morse curves for lower singlet states of an aromatic molecule illustrating the origin of the mirror-image relationship between absorption and fluorescence spectra.

Strong absorption bands are associated therefore with short excited state lifetimes and vice versa, unless non-radiative processes incur. The dye molecule referred to above with $f = 1$ and $\nu_{max} = 500$ nm $= 20\,000$ cm$^{-1}$ has, according to (1.13), $\tau = 3{\cdot}8 \times 10^{-9}$ s.

While the processes of absorption from the ground state $S_0$ of a molecule to a series of excited states $S_1, S_2$, etc., without change of multiplicity are easily understood, the emission from a molecule is more complicated. In addition to the possibility of returning spontaneously to the ground state $S_0$ by radiative emission of light of frequency similar to that characterizing absorption, a process known as *fluorescence* and characterized by exponential kinetics, a molecule can become collisionally deactivated (to give a reduced quantum yield of fluorescence) or it may cross to a triplet level (Fig. 2.4). Emission during return from the latter level to $S_0$ occurs at a longer wavelength than fluorescence and is called *phosphorescence*; it involves a change in multiplicity of $\Delta S = 1$ and is much slower, being of the order of ms, but is still first-order in character. (A fuller discussion of these radiative pathways is given in section 2.4.) The longevity of the $T_1$ state means that the corresponding absorption $T_1 \leftarrow S_0$ has an exceedingly low $f$-value (of the order of $10^{-5}$) with $\epsilon_{max} \sim 0{\cdot}05\ l\ \text{mol}^{-1}\text{cm}^{-1}$. Mean lifetimes of molecules calculated from (1.13) are in good agreement with those obtained experimentally (section 5.5).

An interesting feature of the fluorescence spectra of aromatics in solution is the mirror-image relationship between their vibrational structure and those of the corresponding absorption spectra (Fig. 1.8). This arises because whilst all absorption processes occur from the $v'' = 0$, level of the $S_0$ state, fast vibrational relaxation in the $S_1$ state once populated brings all molecules in that state to the $v' = 0$ level from which they fluoresce (Fig. 1.9). The 0–0 bands are non-coincident because of differential solvation of the two states. The approximately equal separation of the lower vibrational levels in the two electronic states is due to the relatively rigid carbon framework.

## 1.5 Spin conservation rules

The rules concerning changes of the total spin angular momentum quantum number $S$ for single molecules have already been noted. A useful extension of these rules can be made for systems involving interaction between (a) an excited molecule and a molecule in its ground state and (b) two excited molecules, either identical or different. This generalization, known as the Wigner spin conservation rule, can be expressed:

> *The spin states of the reactants can be combined in any way to give those of the products provided no change of the total spin is made.*

Alternatively one can think of maintaining a 'spin balance' just as one normally maintains charge and material balance in stoichiometric equations. The various possibilities according to Wigner's rule become;

| Type of interaction | | Multiplicities of products | |
|---|---|---|---|
| Singlet–singlet $(\uparrow\downarrow)^* + (\uparrow\downarrow)$ | → | $(\uparrow\downarrow) + (\uparrow\downarrow)^*$ | i.e., singlet + singlet* |
| | | $\uparrow\downarrow\uparrow + \uparrow$ | i.e., doublet + doublet |
| | | $\uparrow\downarrow + \uparrow + \downarrow$ | i.e., singlet + doublet + doublet |
| Triplet–singlet $(\uparrow\uparrow) + (\uparrow\downarrow)$ | → | $\uparrow\downarrow + \uparrow\uparrow$ | i.e., singlet + triplet |
| | | $\uparrow\uparrow\downarrow + \uparrow$ | i.e., doublet + doublet |
| | | $\uparrow\uparrow + \downarrow + \uparrow$ | i.e., triplet + doublet + doublet |
| | | $\uparrow + \uparrow + \uparrow\downarrow$ | i.e., doublet + doublet + singlet |
| | | $\uparrow\uparrow\downarrow\uparrow$ | i.e., triplet |
| Triplet–triplet (a) $(\uparrow\uparrow) + (\uparrow\uparrow)$ | → | $\uparrow\uparrow + \uparrow + \uparrow$ | i.e., triplet + doublet + doublet |
| | | $\uparrow\uparrow\uparrow + \uparrow$ | i.e., quartet + doublet |
| | | $\uparrow\uparrow\uparrow\uparrow$ | i.e., quintet |
| (b) $(\uparrow\uparrow) + (\downarrow\downarrow)$ | → | $\uparrow\uparrow\downarrow + \downarrow$ | i.e., doublet + doublet |
| | | $\uparrow\downarrow + \uparrow\downarrow$ | i.e., singlet + singlet |

Some of these possibilities are more often found than others. Plenty of examples of the more universal types will be found in succeeding chapters; it is worth noting at this point, however, that these allowed processes are a key to rationalizing processes such as energy transfer and triplet–triplet annihilation.

Some reassessment of the role of spin in photochemistry will probably need to be made, however, in view of the recent formulation of 'spin-free' theoretical treatments due to Matsen.[8]

## References


1. J. Fritzche, *Z. Angew. Chem.*, **10**, 290 (1867).
2. G. Ciamician and P. Silber, *Chem. Ber.*, **35**, 4128 (1902).
3. A. Mustafa, *Chem. Revs.*, **51**, 1 (1952) and references therein.
4. J. Draper, *Phil. Mag.*, **25**, 1 (1844).
5. W. A. Noyes, Jr. and P. A. Leighton, *The Photochemistry of Gases*, Reinhold, New York, chapter 6 (1941).
6. D. H. O. John and G. T. J. Field, *A Textbook of Photographic Chemistry*, Chapman and Hall, London, chapter 1 (1963).
7. C. A. Parker and W. T. Rees, *Analyst*, **85**, 587 (1960).
8. F. A. Matsen, *J. Amer. Chem. Soc.*, **92**, 3525 (1970).

# 2. Primary photochemical acts

## 2.1 Excitation of atoms in the gas phase

As in the case of spectroscopy, the study of atoms provides a model for the discussion of molecules. Excluding the trivial case of ionization, the remaining possibilities for a photoexcited atom are

(*a*) return to the ground by re-emission of the light energy as fluorescence,
(*b*) crossing to some other electronic state,
(*c*) transfer of the energy to another atom, either of the same or a different type, or to a molecule,
(*d*) addition to a molecule, possibly only for a short time and
(*e*) absorption of a second quantum;

(*c*), (*d*), and (*e*) may, of course, follow (*b*).

Mercury vapour provides the example *par excellence* of atomic photochemistry and the relevant electronic states are depicted in Fig. 2.1. It absorbs light at 184·9 and 253·7 nm to produce 6 ($^1P_1$) and 6 ($^3P_1$) states respectively; the latter is an example of a 'forbidden' $\Delta S > 0$ transition commonly found with heavier elements. The $^1P_1$ state can re-emit 184·9 nm light (fluorescence) or engage in physical or chemical quenching processes. The $^3P_1$ state can re-emit 253·7 nm light (phosphorescence) or become *totally* de-energized by transferring its energy to a second body which may be

(*a*) an atom or molecule capable of dissipating its newly acquired energy without chemical effect, i.e., as kinetic or vibrational energy,
(*b*) a molecule with stable (i.e., non-dissociative) excited states which may re-emit light or react chemically with some further species,
(*c*) a molecule which promptly dissociates on being energized or
(*d*) a molecule which undergoes either isomerization or elimination.

Again the $^3P_1$ state may add to a molecule to give a transient mercury compound which decomposes to give Hg($^1S_0$) and dissociation products. Finally *partial* quenching by a few specific molecules, e.g., $N_2$, $H_2O$, and CO, may produce the

$^3P_0$ state of slightly lower energy, which can undergo all the possible reactions outlined for the $^3P_1$ state, including phosphorescence ($\lambda_P$ = 265·4 nm), but is generally rather less reactive.[1]

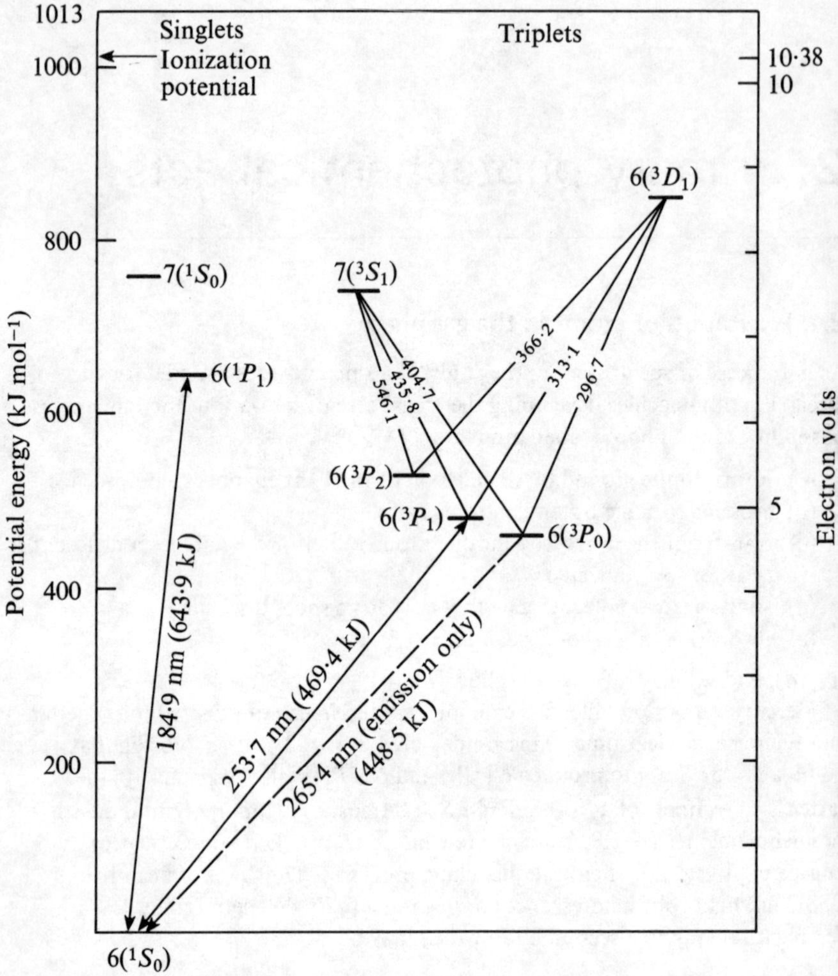


**Fig. 2.1** Lower excited states of the mercury atom. The 184·9 nm and 253·7 nm lines are resonance radiation and appear in absorption from and emission to the ground state, while the 265·4 nm appears only in emission.

The lifetime $\tau(=k_1^{-1})$ of the $^1P_1$ state, which is typical of excited singlet states, is 1·3 ns but that of the $^3P_1$ state, which must return to the singlet ground state via a $\Delta S > 0$ transition, is 110 ns.[2] At low vapour pressures and in the absence of added gases $\varphi_P$ for phosphorescence using 253·7 nm light is 1·0; at higher pressures or in the presence of added gases $\varphi_P$ falls because of the increased probability of alternative reactions. Spontaneous return of the $^3P_0$ state to the ground state violates both the $\Delta S$ and $\Delta J$ selection rules and this state is even longer lived ($\tau \sim 10^6$ ns).[3]

Accounts of mercury-sensitized reactions are legion[4] and we shall confine discussion to a few 'model' systems.

## *Sensitized fluorescence*

Irradiation of mixtures of mercury and thallium vapours with 253·7 nm light, which is absorbed only by the mercury, results in emission from states of thallium atoms excited by energy transfer from Hg $6(^3P_1)$.[5] This state is capable of sensitizing fluorescence from a variety of metal atoms provided the two electronic states involved are not greatly disparate in energy (i.e., $\not>$ 60 kJ). This type of process is given further consideration in section 3.1 on energy transfer.

## *Sensitized photolysis*

Molecules completely transparent to 253·7 nm light can be decomposed by $Hg(^3P_1)$ atoms, the simplest example being hydrogen:

$$Hg(^3P_1) + H_2 \rightarrow Hg(^1S_0) + 2H^{\bullet}(^2S_{1/2})$$

$Hg(^1P_1)$ atoms are also capable of sensitizing decompositions and a comparison of the behaviour of these two atomic sensitizers of different multiplicity gives a good illustration of the spin conservation rule (p. 21); for example, the net reactions with alkanes and alkenes, RH, in their ground singlet states, $^1\Sigma$, can be summarized;[6]

$$Hg(^1P_1) + RH(^1\Sigma) \rightarrow \begin{cases} Hg(^1S_0) + RH(^1\Sigma)^* & (2.1) \\ Hg(^1S_0) + RH(^1\Sigma)^v & (2.2) \\ Hg(^1S_0) + R^{\bullet}(^2\Sigma) + H^{\bullet}(^2S) & (2.3) \\ HgH^{\bullet}(^2\Sigma) + R^{\bullet}(^2\Sigma) & (2.4) \end{cases}$$

$$Hg(^3P_1) + RH(^1\Sigma) \rightarrow \begin{cases} Hg(^1S_0) + RH(^3\Sigma) & (2.5) \\ Hg(^1S_0) + R^{\bullet}(^2\Sigma) + H^{\bullet}(^2S) & (2.6) \\ HgH^{\bullet}(^2\Sigma) + R^{\bullet}(^2\Sigma) & (2.7) \end{cases}$$

For alkanes the level of the $^1\Sigma^*$ state is too high for reaction 2.1 to be possible. Irradiation of mercury–alkane mixtures with 184·9 nm light appears to proceed mostly through reactions 2.3 and and 2.4. Simple alkenes, on the other hand, have their first $^1\Sigma^*$ state just below the 184·9 nm level and reaction 2.1 is energetically possible and appears to be realized in experiment.

The reaction scheme suggests that reaction on irradiation with 253·7 nm light should proceed via steps 2.5 to 2.7; however, 2.7 is eliminated on energetic grounds for alkanes but remains feasible for alkenes.

Once certain possible reaction paths have been ruled out on energetic grounds, we can attempt to relate the remaining possibilities to the observed products and their quantum yields. At *low* conversions $Hg(^3P_1)$ reaction with alkanes produces $H_2$ with $\varphi_{H_2}$ approaching unity; at higher degrees of conversion, $\varphi_{H_2}$ falls to 0·6–0·7, indicating scavenging of H atoms by secondary products such as alkenes. If one primary product from RH is $H^{\bullet}$ then the other should be $R^{\bullet}$ and mass-spectrometric and 'trapping' experiments show this to be the case. Two main mechanisms are currently under discussion[7] typified by reactions 2.8 and 2.9:

$$RCH_2CH_3 + Hg(^3P_1) \rightarrow \begin{cases} R\dot{C}HCH_3 + H^{\bullet} + Hg(^1S_0) \\ RCH_2CH_2^{\bullet} + H^{\bullet} + Hg(^1S_0) \end{cases} \qquad (2.8)$$

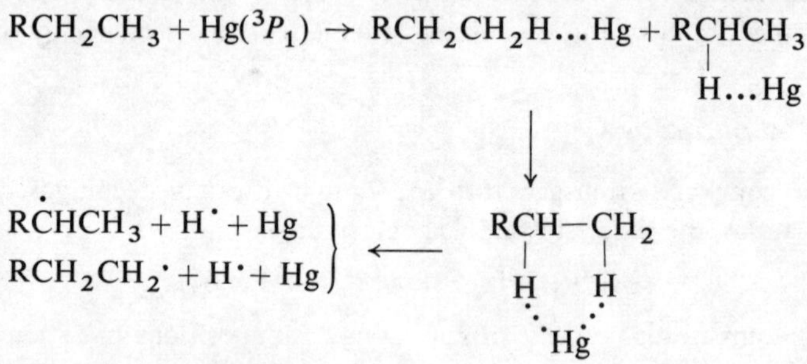

(2.9)

The $Hg(^3P_1)$-sensitized decomposition of ethylene produces $H_2$ and $C_2H_2$ each with a $\varphi$ value of about 0·4. The origin of the $H_2$ is not atomic because of the failure of sensitized mixtures of $C_2D_4$ and $C_2H_4$ to produce significant amounts of HD or $C_2HD$. The decrease of $\varphi$ as the pressure is increased or an inert gas added suggests deactivation of an excited state which otherwise undergoes a molecular elimination of $H_2$. Further detail is supplied by experiments[8] with *cis*-ethylene-$d_2$ which produces, in addition to $D_2$, $H_2$, HD, and acetylene, the isomers *trans*-ethylene-$d_2$ and $H_2C{=}CD_2$. These results are tentatively rationalized in the scheme where $^3(C_2H_4)^{v'}$ may be vibrationally excited triplet ethylidene,

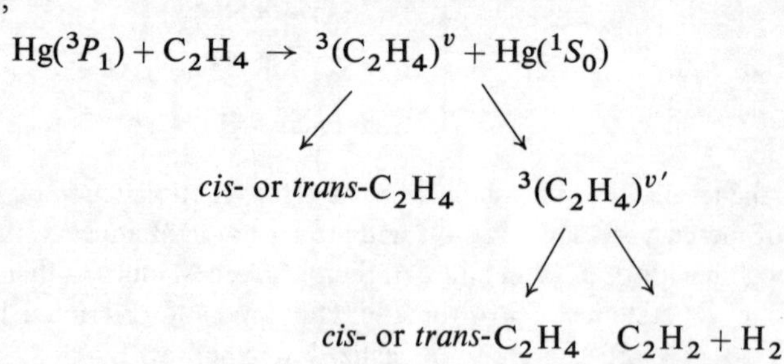

*cis*-But-2-ene undergoes isomerization ultimately to an equimolar mixture of *cis*- and *trans*-isomers on reaction with $Hg(^3P_1)$, much as it does following energy transfer from triplet benzene ($^3B_{1u}$).

Other examples of $Hg(^3P_1)$-photosensitized reactions are given in Table 2.1. The primary path given is not necessarily exclusive.

Closely analogous to $Hg(^3P_1)$ sensitizations are those of $Cd(^3P_1)$ formed by 326·1 nm light. The rather different reactions induced by atomic oxygen do not come into the present category of those initiated following light absorption by atoms.

TABLE 2.1: $Hg(^3P_1)$ sensitized reactions

| Substrate | Primary products |
|---|---|
| $CH_3OH$ | $CH_3O^{\bullet} + H^{\bullet} + Hg(^1S_0)$ |
| $CH_3COCH_3$ | $CH_3^{\bullet} + CH_3CO^{\bullet} + Hg(^1S_0)$ |
| $CH_3CHO$ | $CH_3^{\bullet} + CHO^{\bullet} + Hg(^1S_0)$ |
| $CH_3CH{=}CHCHO$ | $CH_3CH{=}CH_2 + CO + Hg(^1S_0)$ |
| $CH_3Cl$ | $CH_3^{\bullet} + Cl^{\bullet} + Hg(^1S_0)$ |
| $N_2O$ | $N_2(^1\Sigma) + O(^3P) + Hg(^1S_0)$ |
| $NH_3$ | $NH_2^{\bullet} + H^{\bullet} + Hg(^1S_0)$ |

## 2.2 Excitation of diatomic molecules

Electronic excitation of diatomics is normally accompanied by a set of vibrational and rotational changes (p. 6). Usually the chemical properties of an excited molecule are determined exclusively by its electronic state although in certain cases the possession of many quanta of vibrational energy can confer a degree of reactivity upon a molecule not apparent in a molecule in the zeroth vibrational level (of the same electronic state). A familiar example of the latter phenomenon is that of predissociation whereby a molecule with sufficient vibrational energy may cross over from a bonding state to a second dissociative state which intersects with the original state. This is illustrated for $S_2$ in Fig. 2.3.

An electronically excited diatomic molecule may undergo all the processes outlined for atoms. A fluorescence spectrum will be exhibited when a minimum exists in the excited state concerned, and it will be complicated by the appearance of bands due to concomitant vibrational changes. Fluorescence studies on simple molecules are normally carried out on gaseous samples, degradation of electronic and vibrational energy to heat occurring more readily in solution during collisions with solvent molecules. The visible emission from molecular iodine vapour is from the $^3\Pi^+_{0u}$ state ($v' = 26$) and is really an example of resonance phosphorescence brought about by the breakdown of the $\Delta S = 0$ selection rule by the heavy atoms. A simplified picture of the Morse curves is given in Fig. 2.2 based on the discussion of Mathieson and Rees.[9]

A possibility naturally peculiar to molecules is that of dissociation. This may occur through a direct excitation from the ground state to a purely dissociative upper state, a process associated with a structureless, continuous absorption. Again, the upper state may exhibit a minimum, but excitation to vibrational levels of progressively higher energy may ultimately lead to dissociation and an associated region of continuous absorption. Crossing over from a bonding upper state to a dissociative state may occur, leading to the phenomenon of pre-dissociation already mentioned.

These various modes of dissociation are illustrated by the cases of $H_2$, HI, and $S_2$ respectively. Ground state $H_2$ exists in a $^1\Sigma_g^+$ state and on excitation with 110·9 and 100·2 nm light enters $^1\Sigma_u^+$ and $^1\Pi_u$ states, corresponding to promotion of a $1s\sigma_g$ electron to $2p\sigma_u^*$ and $2p\pi_u^*$ MO's respectively; the operation

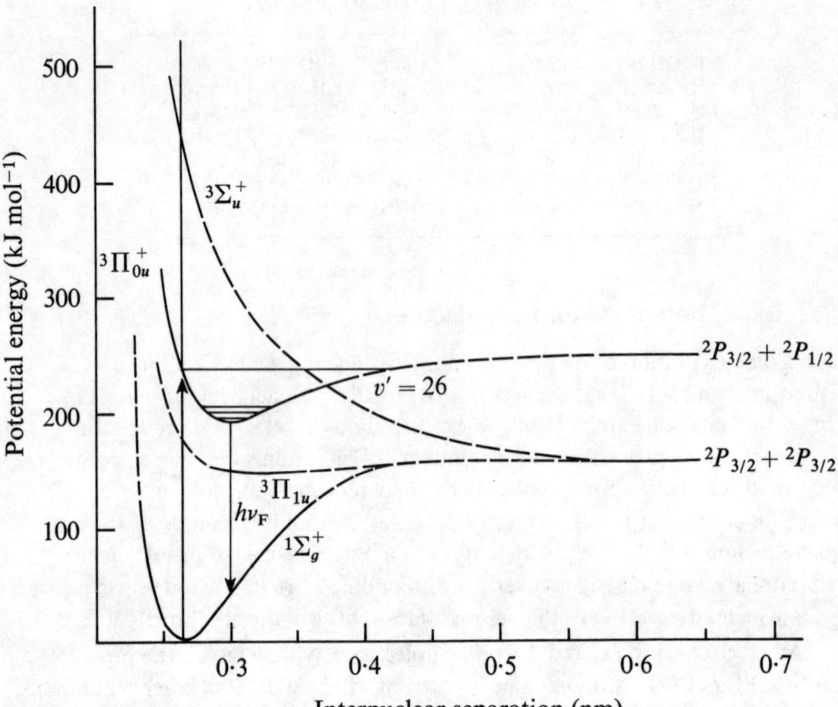


**Fig. 2.2** Morse curves for $I_2$. (Reproduced with permission of the authors and publisher.[9])

of selection rules (p. 4) prevents other types of promotion. Both of these singlet states are bonding, but absorption of light of progressively larger energies leads to the dissociation of each of them to a ground state hydrogen atom and an excited hydrogen atom ($n = 2$). With both of these singlet states, however, are associated triplet states, populated by inverting one electron spin to give the transitions $^3\Sigma_u^+ \leftarrow {^1\Sigma_u^+}$ and $^3\Pi_u \leftarrow {^1\Pi_u}$. The $^3\Sigma_u^+$ state is purely dissociative, giving *two* ground state hydrogen atoms, the excess electronic energy appearing as kinetic energy.

The absorption spectrum of gaseous HI is continuous throughout and *no* bonding upper states exist. The $^3\Pi_1$ and $^1\Pi$ states yield ground state H and

$I(^2P_{3/2})$ atoms and the $^3\Pi_0^+$ state yields a ground state H atom and an excited $I(^2P_{1/2})$ atom. The dissipation of electronic energy in the form of translation energy confers upon the H atom exceptional reactivity and it is designated a 'hot' atom. The quantum yield for loss of HI is 2·0, suggesting a mechanism:

$$HI \xrightarrow{h\nu} H^{\bullet} + I^{\bullet}$$

$$H^{\bullet} + HI \longrightarrow H_2 + I^{\bullet}$$

$$2I^{\bullet} + M \longrightarrow I_2 + M$$

$S_2$, like $O_2$, exists in a ground triplet state for similar reasons. The first transition is $^3\Sigma_u^- \leftarrow {}^3\Sigma_g^-$, the upper state exhibiting a minimum in its Morse curve (Fig. 2.3).[10] Excitation up to the $v' = 9$ level in the $^3\Sigma_u^-$ state is accompanied by the expected sharp rotational fine structure. Excitation to the

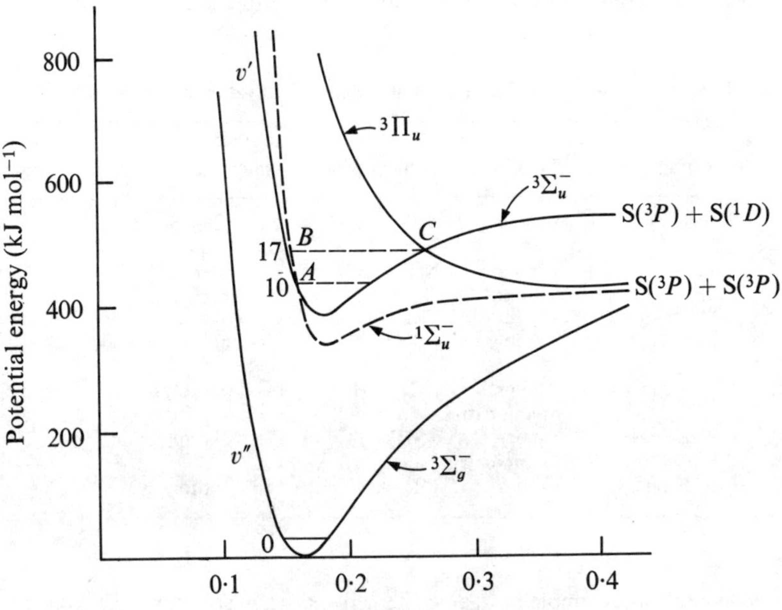


Fig. 2.3 Morse curves for $S_2$, after Bowen.[10] (Reproduced with permission of the author and publisher.)

levels $v' = 10$ to 17 results in a blurring of this fine structure because in this region intersystem crossing to the $^1\Sigma_u^-$ state occurs. Excitation to even higher vibrational levels results in crossing to the repulsive $^3\Pi_u$ state which produces two sulphur ($^3P$) atoms.

## 2.3 Excitation of simple polyatomic molecules

The types of transition undergone by these molecules closely parallel those discussed for diatomic molecules, and the photolysis products are normally either an atom plus a small free radical, or two radicals, either of which or both may be in an excited state. Vacuum u.v. irradiation may also induce photoionizatio The exact distribution of electronic energy among the products is a function of the wavelength of the exciting light, for example, with $H_2O$,

$$H_2O \xrightarrow{\lambda < 242\ \text{nm}} H^{\bullet}(^2S_{1/2}) + OH^{\bullet}(^2\Pi)$$

$$H_2O \xrightarrow{\lambda < 135{\cdot}6\ \text{nm}} H^{\bullet}(^2S_{1/2}) + OH^{\bullet}(^2\Sigma^+)$$

The effect of phase is also important; for 253·7 nm irradiation of $H_2O_2$, $\varphi(-H_2O_2) = 0{\cdot}85$ in the gas phase but in solution $\varphi(OH^{\bullet}) \sim 1$, i.e., $\varphi(-H_2O_2) \sim 0{\cdot}5$ This illustrates the enhanced chances of recombination of the primary radicals in a solvent cage.

Other examples[11-19] of primary acts are given in Table 2.2.

TABLE 2.2: Chief primary photochemical acts for simple polyatomic molecules

| Molecule | λ (nm) | Phase | Primary process | Method of examination | Reference |
|---|---|---|---|---|---|
| $H_2S$ | 185 | Vapour | $SH^{\bullet} + (H^{\bullet})^*$ | Scavenging of $(H^{\bullet})^*$ by $C_4D_{10}$ | 11 |
| $NO_2^{\bullet}$ | 270–360 | Vapour | $NO^{\bullet} + O(^3P)$ | Manometrically | 12 |
| NOCl | 253·7–635 | Vapour | $NO^{\bullet} + Cl^{\bullet}$ | Manometrically | 13 |
| $ClO_2^{\bullet}$ | Wide range | Vapour | $ClO^{\bullet} + O(^3P)$ | Flash photolysis | 14 |
| $O_3$ | Wide range | Vapour | $O_2 + O(^1D)$ | Flash photolysis and detection of $(OH^{\bullet})^{v}$ | 15 |
| $NH_3$ | Wide range | Vapour | $NH_2^{\bullet} + H^{\bullet}$ | Flash photolysis | 16 |
| | | Argon matrix | | E.s.r. | 17 |
| $N_2H_4$ | Wide range | Vapour | $2NH_2^{\bullet}$ | Flash photolysis | 18 |
| | | Argon matrix | | E.s.r. | 17 |
| $PH_3$ | Wide range | Vapour | $PH_2^{\bullet} + H^{\bullet}$ | Flash photolysis | 19 |

The excitation of simple inorganic charged species in solution, e.g., iodide and ceric ions, frequently involves participation by the electrons of the molecules comprising the first solvation shell and in this sense the system should be regarded as complex, and as such is discussed in section 2.5.2.

The production of simple radicals in primary photochemical acts implied by kinetic studies[4,20] has been confirmed spectroscopically in a number of ways (Table 2.2). The radical may be detected optically in emission, e.g., in the 306·2 nm 0–0 fluorescence band of the $OH^{\bullet}(^2\Sigma^+) \rightarrow OH^{\bullet}(^2\Pi)$ transition or in absorption, e.g., in the 450–740 nm band system of $NH_2^{\bullet}$ following flash photolysis of gaseous ammonia.[16]

Again the molecule may be subjected to photolysis in dilute solution in an inert gas or glassy alcohol matrix at liquid helium or nitrogen temperatures and its i.r., u.v., and e.s.r. spectra taken; for example, the photolysis of dilute solutions of $H_2O$, $N_2H_4$, $SiH_4$, and HCN in argon at 4 K produce e.s.r. spectra attributed to $H^\bullet$, $NH_2^\bullet$, $SiH_3^\bullet$, and $CN^\bullet$ respectively.[17]

The presence and yield of small radicals can be adduced from their reactions, e.g., with reactive 'scavengers' or monomers capable of undergoing polymerization both in the gas phase and in solution (section 3.2.5).

## 2.4 Excitation of complex polyatomic molecules

The behaviour of complex molecules on photolysis falls into several well-defined classes. Certain complex molecules behave exactly as their more simple analogues. If, for example, the molecule consists of a number of relatively strong bonds and one or two weak bonds, and if the electrons comprising these bonds are essentially non-interacting, then homolytic fission will occur especially at relatively long wavelengths of irradiation, e.g.,

$$(CH_3)_3C{-}O{-}O{-}C(CH_3)_3 \xrightarrow[253{\cdot}7\ \text{nm}]{h\nu} 2(CH_3)_3C{-}O^{\bullet *} \quad \varphi = 1{\cdot}0 \tag{2.10}$$

Subsequent fragmentation of hot radicals may occur, particularly at low gas pressures when the possibilities of collisional deactivation are limited, e.g.,

$$(CH_3)_3C{-}O^{\bullet *} \rightarrow CH_3COCH_3 + CH_3^\bullet$$

At shorter wavelengths a more endothermic process may be favoured,

$$(CH_3)_3C{-}O{-}O{-}C(CH_3)_3 \xrightarrow[190\text{–}330\ \text{nm}]{h\nu} (CH_3)_3C{-}O{-}O^\bullet + C(CH_3)_3^\bullet$$

Quantum yield and kinetic information[4,20] is again augmented by the application of flash and e.s.r. spectroscopy. Vapour phase flash photolysis of formaldehyde, acetaldehyde, and glyoxal produces the absorption spectrum of $HCO^\bullet$ in each case; however, $\varphi(HCO^\bullet)$ is very different for each example and is dependent on the wavelength of irradiation, photoelimination of CO providing an important alternative path at shorter wavelengths ($\lambda < 300$ nm) for many carbonyl compounds (section 4.1.1).

Aliphatic compounds may undergo radical fragmentation simultaneously with extrusion of a thermodynamically highly stable molecule such as $N_2$, e.g.,

$$(CH_3)_2\underset{\substack{|\\ CN}}{C}{-}N{=}N{-}\underset{\substack{|\\ CN}}{C}(CH_3)_2 \xrightarrow[366\ \text{nm}]{h\nu} 2(CH_3)_2\underset{\substack{|\\ CN}}{C^\bullet} + N_2 \quad \varphi = 0{\cdot}43 \tag{2.11}$$

At short wavelengths even alkanes can undergo homolysis, although extrusion of $H_2$ to yield a carbene remains the dominant pathway until at very low wavelengths (97 nm for $CH_4$) photoionization becomes possible, being exclusive at $\lambda < 80$ nm.

$$CH_4 + h\nu \rightarrow \begin{cases} CH_2\!:\, + H_2 & \varphi = 0{\cdot}39 \\ CH_3^{\bullet} + H^{\bullet} & \varphi \sim 0{\cdot}07 \\ \overset{+}{CH_4}\!\dot{\ }\, + e^- & \varphi = 1{\cdot}0 \quad (30\text{–}80 \text{ nm}) \end{cases}$$

Further examples of the dominant primary step in photolysis of saturated molecules are given in Table 2.3.[21-29]

The presence of a $\pi$-bond in a large molecule immediately introduces fresh possibilities, in particular long-wavelength $\pi^* \leftarrow \pi$ or $\pi^* \leftarrow n$ transitions to give excited states of sufficiently low energy *not* to fragment but rather to undergo abstraction or cyclization reactions. This situation is obviously an important feature of organic photochemistry, particularly with regard to that of conjugated chain and cyclic hydrocarbons and ketones. Associated with this longest-wavelength transition (which may be quite strong even when forbidden on symmetry grounds) will be the lowest excited singlet state of the molecule (denoted in the general case by $S_1$) and related to this state will be that entered following inversion of a single electron spin, i.e., the first triplet state $T_1$. There will also exist higher energy states, $S_2$ and $T_2$, etc., with an inter-relation apparent from a typical state or 'Jablonski' diagram in which are included the vibrational and rotational levels (Fig. 2.4). Consider the possibilities available for the

TABLE 2.3: Chief primary photochemical acts for saturated organic molecules

| Molecule | λ (nm) | Phase | Primary process | Reference |
|---|---|---|---|---|
| $CH_3OH$ | 180–200 | Vapour | $CH_2O + H_2$ | 21 |
| $CH_2$—$CH_2$ \ / O | < 200 | Vapour | $CH_3^{\bullet} + HCO^{\bullet}$ | 22 |
| $CH_3OOCH_3$ | 253·7 | Vapour | $2(CH_3O^{\bullet})^*$ or $2(CH_2O + H^{\bullet})$ | 23 |
| $(CH_3)_3COOH$ | 313·0 | Inert solvent | $(CH_3)_3CO^{\bullet} + OH^{\bullet}$ | 24 |
| $CH_3NH_2$ | 194–234·5 | Vapour | $CH_3NH^{\bullet} + H^{\bullet}$ | 25 |
| $CH_3SH$ | 253·7 | Vapour | $CH_3S^{\bullet} + H^{\bullet}$ | 26 |
| $C_2H_5I$ | 253·7 | Vapour | $(C_2H_5^{\bullet})^* + I^{\bullet}$ | 27 |
| | | Film (77 K) | $C_2H_4 + HI$ | 28 |
| $CH_3Br$ | 253·7 | Vapour | $CH_3^{\bullet} + Br^{\bullet}$ | 29 |

'typical' molecule following a normal absorption of a quantum of light of sufficient energy to place it in an excited vibrational level of its first excited singlet state, i.e., $(S_1)^v$:

(*a*) the molecule will lose excess vibrational energy by collisions until it reaches the $v = 0$ level of $S_1$ (a process denoted vibrational relaxation (VR) or cascade). It may then fluoresce by emitting a quantum of light $h\nu_F$, terminating in an excited vibrational level of the state $S_0$. The origin of the fluorescence spectrum is considered in section 1.4. The molecule will lose further excess vibrational energy until it achieves thermal equilibrium with its surroundings.

(*b*) On reaching the state $S_1(v = 0)$ the molecule may cross over into a very high vibrational level of the intersecting $S_0$ state—a process designated

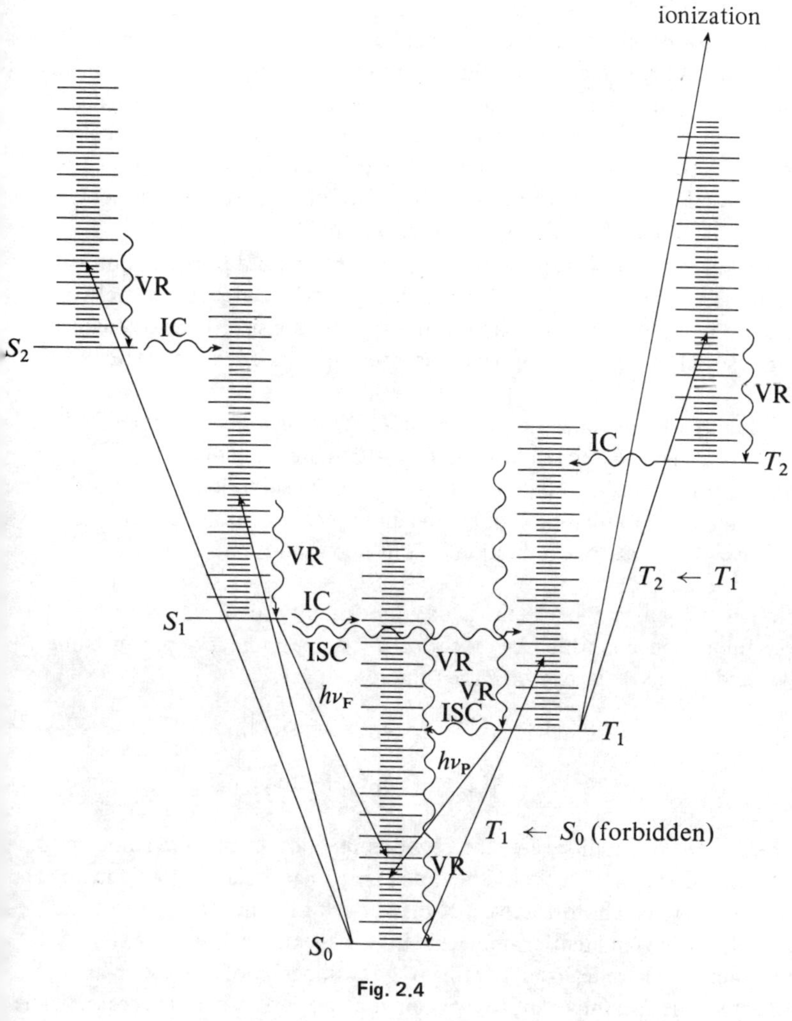


Fig. 2.4

internal conversion (IC). The many quanta of vibrational energy are lost in successive collisions until, again, the molecule reaches thermal equilibrium. The result of this sequence of steps is the degradation of electronic energy to heat.

(c) The molecule may cross from the $S_1(v = 0)$ level to an excited vibrational level of the first triplet state $(T_1)^v$. This violates the $\Delta S = 0$ rule but, as will be seen, it is a common occurrence and is known as inter-system crossing (ISC). The molecule then loses vibrational energy to reach the $T_1(v = 0)$ state following which it re-enters the $S_0$ state *either* by emitting a quantum of phosphorescence $h\nu_P$ *or* by a second act of inter-system crossing to give $(S_0)^v$.

A number of other possibilities are encountered either with peculiar molecules or under special experimental conditions. Azulene fluoresces only from the $S_2$ (and not the $S_1$) state (further information on azulene is given in section 5.5). The forbidden $T_1 \leftarrow S_0$ transition can be realized for a number of molecules *either* by using an intense light source such as a laser or by promoting a breakdown of the $\Delta S = 0$ rule by introducing heavy atoms or paramagnetic centres either into the molecule or the solvent medium (p. 39).

Following the introduction firstly of $\mu$s and then ns and ps flash photolysis, the excitations from the $S_1$ and $T_1$ states to $S_2$ and $T_2$ states respectively can be achieved by absorption of quanta of the appropriate size. This has enabled the compilation of a series of optical absorption spectra of excited states (section 5.5).

Molecules with low ionization potentials (I.P.) in their ground states $S_0$ have even lower I.P.'s when excited to their $T_1(v = 0)$ states and absorption of a second quantum of light of only moderate size may lead to ionization. Such a process is termed two-quantum ionization and is characterized by a *squared* dependence of the rate of production of ions upon the incident light intensity.[30]

Another energetically- and spin-'allowed' possibility for $T_1$ states is that of delayed fluorescence in which *two* identical $T_1$ states interact to produce one $S_0$ state and one $(S_1)^v$ state which then fluoresces;

$$T_1(\uparrow\uparrow) + T_1(\uparrow\uparrow) \rightarrow S_0(\downarrow\uparrow) + S_1(\downarrow\uparrow) \tag{2.12}$$

$$S_1(\downarrow\uparrow) \rightarrow S_0(\downarrow\uparrow) + h\nu_F \tag{2.13}$$

These various possibilities are all realized in practice, but before going on to discuss selected examples it should be noted that a particular molecule in an excited state may be photoreactive and may attack the solvent or some other molecule, including an identical molecule in its $S_0$ state, and to this extent the picture given is rather idealized, relating to a few model molecules such as naphthalene. It is also important to gain an appreciation of the time-scales of the transitions: these are normally expressed in terms of

(a) the half-life in seconds, $t_{1/2}$, for the transition, state (2) → state (1),
(b) the rate constant $k_1$ for the process ($k_1 = 0{\cdot}693/t_{1/2}$) or,
(c) the corresponding relaxation time $\tau$, i.e., the time for 1/e of the molecules in state (1) to reach state (2); also $\tau = k_1^{-1}$.

Typical values for $t_{1/2}$ are given:

| Process | $t_{1/2}$ (s) |
|---|---|
| Fluorescence ($S \rightarrow S_0$) | $10^{-9}$–$10^{-8}$ |
| ISC ($S_1 \rightarrow T_1{}^v$) | $10^{-9}$–$10^{-8}$ |
| Phosphorescence ($T_1 \rightarrow S_0$) | $10^{-3}$–1 |
| Delayed fluorescence (fluids) | $10^{-5}$–$10^{-4}$ |

The comparative longevity of the $T_1$ state accords with the selection rule $\Delta S = 0$ and partly accounts for its central significance in photochemistry. The slowness of delayed fluorescence is due to the necessity of two triplet states to diffuse within ~ 1 nm for exchange of spins to occur. $t_{1/2}$ for intersystem crossing is based on the observation that in many cases this pathway competes very effectively with fluorescence even when the latter is fast.

Ignoring for the moment the more esoteric of these processes, we can summarize the position for a 'typical' molecule following excitation to $S_1$ by

$$\varphi = \varphi_F + \varphi_T + \varphi_{IC}$$

i.e., the absorption of one quantum must result in some (or nil) fluorescence, triplet state formation (not necessarily equivalent to phosphorescence, $\varphi_P$, but $\varphi_P \leqslant \varphi_T$), and internal conversion. $\varphi_F$ and $\varphi_T$ have been determined for numerous molecules and the selection of results presented in Table 2.4 indicates the general magnitudes of these parameters.

TABLE 2.4: Fluorescence and triplet state quantum yields for some polynuclear aromatic molecules in solution

| Molecule | $\varphi_F$ | $\varphi_T$ |
|---|---|---|
| Naphthalene | 0·21 | 0·71 |
| Anthracene | 0·33 | 0·58 ± 0·10 |
| Phenanthrene | 0·14 | 0·70 ± 0·12 |
| Triphenylene | 0·09 | 0·89 |
| 1,2,5,6-Dibenzanthracene | – | 1·03 ± 0·16 |
| Pyrene (< $10^{-5}$ M) | 0·65 | 0·38 |
| 1-Methoxynaphthalene | 0·53 | 0·46 |
| 9-Phenylanthracene | 0·45 | 0·505 ± 0·025 |
| Fluorescein | 0·92 | 0·05 ± 0·02 |
| Dibromofluorescein | – | 0·49 ± 0·07 |
| Eosin | 0·19 | 0·71 ± 0·10 |
| Erythrosin | 0·02 | 1·07 ± 0·03 |

It is clear firstly that $\varphi_F + \varphi_T \simeq 1{\cdot}0$ for all the systems given (where complete information is available). This raises the question of the general importance of the internal conversion $S_1 \rightarrow (S_0)^v$ and it appears that $\varphi_{IC}$ for this particular non-radiative transition is insignificant for the majority of polyacenes. Also of interest is the effect of substitution by heavy atoms of increasing $\varphi_T$ at the expense of $\varphi_F$ in the series of fluoresceins.[32]

Turning now to the transitions

$$T_1 \rightarrow S_0 + h\nu_P \quad \text{(phosphorescence)}$$
$$T_1 \rightarrow (S_0)^v \quad \text{(intersystem crossing)}$$
$$T_1 + T_1 \rightarrow S_1 + S_0 \quad \text{(delayed fluorescence)}$$

it is evident that in the absence of chemical reaction

$$\varphi_T = \varphi_P + \varphi_{ISC} + \varphi_{DF}$$

The conditions for a molecule to phosphoresce, even when $\varphi_T$ is high, are critical. Significant phosphorescence does not normally occur at room temperature in fluid solution (although biacetyl is an interesting exception to this rule) because, (*a*) the excited molecule can encounter quenching impurities such as $O_2$ during its comparatively long lifetime and, (*b*) triplet–triplet annihilation and collisional deactivation may also be effective, although the present evidence indicates (*a*) to be more important than (*b*). It appears that the use of fully fluorinated solvents may also promote phosphorescence in molecules like benzophenone which can abstract H atoms even from benzene.[33] Quenching processes are eliminated on cooling a solution of the molecule in a glassy solvent to 77 K or 4 K or by preparing a solution in a rigid plastic or glass at room temperature, and under these conditions phosphorescence spectra and lifetimes have been measured for numerous molecules and representative results are shown in Table 2.5.

Certain features are immediately apparent, in particular the natural segregation of hydrocarbons and carbonyl compounds on the basis of $\varphi_P$, $\varphi_P/\varphi_F$, and $\tau_P$ values. The effect of introducing heavy atoms, which we have seen to increase $\varphi_P$ at the expense of $\varphi_F$ in Table 2.4, increases $\varphi_P$ and also results in a drastic

TABLE 2.5: Phosphorescence yields, wavelengths, and lifetimes for solutions of aromatics at 77 K

| Molecule | $\varphi_P$ | $\lambda_P$ (nm) | $\tau_P$ (s) | $\varphi_P/\varphi_F$ |
|---|---|---|---|---|
| Benzene | 0·23 | 340 | 7·0 | 0·98 |
| Naphthalene | 0·03 | 470 | 2·3 | 0·09 |
| Naphthalene-$d^8$ | 0·06 | 467 | 9·5 | 0·21 |
| 1-Methylnaphthalene | 0·023 | 476 | 2·1 | 0·05 |
| 1-Chloronaphthalene | 0·16 | 483 | 0·20 | 5·2 |
| 1-Bromonaphthalene | 0·14 | 484 | 0·018 | 164 |
| 1-Iodonaphthalene | 0·20 | 488 | 0·0020 | 1000 |
| Triphenylene | 0·42 | 420 | 15·9 | 2·8 |
| Phenanthrene | 0·135 | 461 | 3·9 | 1·1 |
| Benzaldehyde | 0·40 | 401 | $1·4 \times 10^{-3}$ | $> 10^3$ |
| Benzophenone | 0·74 | 412 | $4·7 \times 10^{-3}$ | $> 10^3$ |
| Acetophenone | 0·62 | 388 | $2·3 \times 10^{-3}$ | $> 10^3$ |
| *m*-Iodobenzaldehyde | 0·64 | 404 | $6·5 \times 10^{-4}$ | $> 10^3$ |

shortening of $\tau_P$ and an increase in the $\varphi_P/\varphi_F$ ratio. Deuteration also increases $\varphi_P$ and $\tau_P$. $\lambda_P$ in all cases is greater than the corresponding $\lambda_F$, which merely reflects the relative positioning of the $S_1$ and $T_1$ levels in the Jablonski diagram. Another general effect is that of the energy gap between the $T_1$ and $S_0$ levels upon $\varphi_P$; for a structurally related set of molecules $\varphi_P$ falls as the gap diminishes.

These experimentally established patterns of behaviour require rationalization. The strongest phosphorescence is found with molecules with a $S_1$ state of $(n, \pi^*)$ character, for example, aldehydes, ketones, *N*-heterocyclic compounds and nitroaromatics. Normally this $\pi^* \leftarrow n$ transition will be of lower energy than the associated $\pi^* \leftarrow \pi$ transition, but in certain cases the energies may be similar or the sequence may be inverted, or even reversed for a *given* molecule by a simple change of solvent. (This also accounts for the fluorescence of chlorophyll in polar solvents, in which the $(\pi, \pi^*)$ state is lower, and the absence of fluorescence in dry hydrocarbon solvents, in which the $(n, \pi^*)$ state is lower.) Again, there will be $^3(n, \pi^*)$ and $^3(\pi, \pi^*)$ levels associated with their singlet counterparts and these may also be 'inverted' in certain molecules. These situations are depicted in Fig. 2.5 which also illustrates the possible relative positioning of the $(n, \pi^*)$ singlet and triplet states.[34]

Situations (a) and (b) are most commonly encountered, i.e., molecules exhibiting a lowest singlet state of $(n, \pi^*)$ type often, but not invariably, possess a lowest triplet $(n, \pi^*)$ state, and will be characterized by low $\varphi_F$ and high $\varphi_P$. This is the case for many carbonyl and nitro compounds and N-heterocycles. The $S_1$-$T_1$ energy gap is also found to be smaller than is the case with polyacenes. Lowest triplet states of $(\pi, \pi^*)$ type are found with phenazine and quinoline with considerable consequences for their basic photochemistry. Classification of a lowest (phosphorescing) triplet level as $(n, \pi^*)$ or $(\pi, \pi^*)$ depends on consideration of several features including,[35]

(*a*) $\tau_P$, which is shorter for $(n, \pi^*)$ than $(\pi, \pi^*)$ states
(*b*) The $S_1$-$T_1$ energy gap which is usually smaller for $(n, \pi^*)$ than $(\pi, \pi^*)$ states,
(*c*) vibrational structure in the phosphorescence spectrum, which may show typical, say, carbonyl frequencies for $(n, \pi^*)$ states, but skeletal frequencies as in polyacene phosphorescence, for $(\pi, \pi^*)$ states,
(*d*) the effect of introducing heavy atoms external to the molecule, which is far greater upon the very weak $(T_1 \leftarrow S_0)$ transitions when these involve a $T_1(\pi, \pi^*)$ as opposed to a $T_1(n, \pi^*)$ state,
(*e*) certain traits of chemical reactivity, for example an $(n, \pi^*)$ triplet state aromatic ketone exhibits stronger powers of hydrogen abstraction, e.g., from the solvent, than an analogue of $(\pi, \pi^*)$ type
(*f*) changing solvent from a hydrocarbon to a polar molecule normally results in a blue shift of an $\pi^* \leftarrow n$ transition and a smaller red shift on a $\pi^* \leftarrow \pi$ transition, both as regards the singlet and triplet manifolds. Such a marked solvent dependence of $\lambda_P$ is diagnostic of an $(n, \pi^*)$ triplet state.

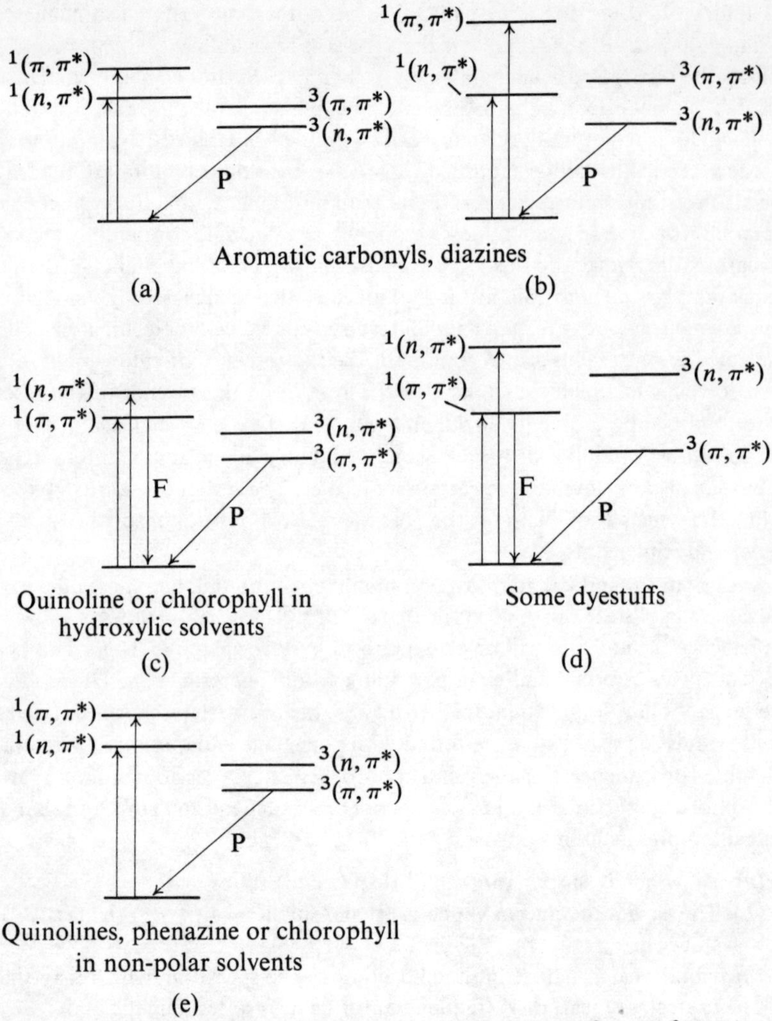


**Fig. 2.5** Significant arrangements of the lower $^1(\pi, \pi^*)$, $^1(n, \pi^*)$ and $^3(\pi, \pi^*)$, $^3(n, \pi^*)$ energ levels. (Reproduced with permission of the authors and publisher.[34])

We have established the inter-relation of $\varphi_F$, $\varphi_T$, $\varphi_P$, and $\varphi_{IC}$ and have correlated these with the relative positioning of $(n, \pi^*)$ and $(\pi, \pi^*)$ states. These considerations apply, however, only to isolated molecules and are invalidated by any one of several perturbing influences. Certain of these, especially interaction with ground state solutes, are discussed in sections 3.1 and 3.2.1. Physical perturbations include

(*a*) internal heavy atom effect,
(*b*) external heavy atom effect,
(*c*) effect of paramagnetic solutes,
(*d*) effect of deuteration,
(*e*) triplet–triplet interaction,
(*f*) absorption of a second quantum.

From Table 2.5 it is evident that effect of substitution of naphthalene by successively larger halogen atoms is to reduce $\tau_P$ and to increase $\varphi_P/\varphi_F$. These imply increases in the rates of the forbidden $T_1 \rightarrow S_0$ and $S_1 \rightarrow T_1$ transitions, which leaves $\varphi_P$ subject to two opposing influences. These effects are quite general and are denoted 'internal heavy atom perturbation'. A further mani-

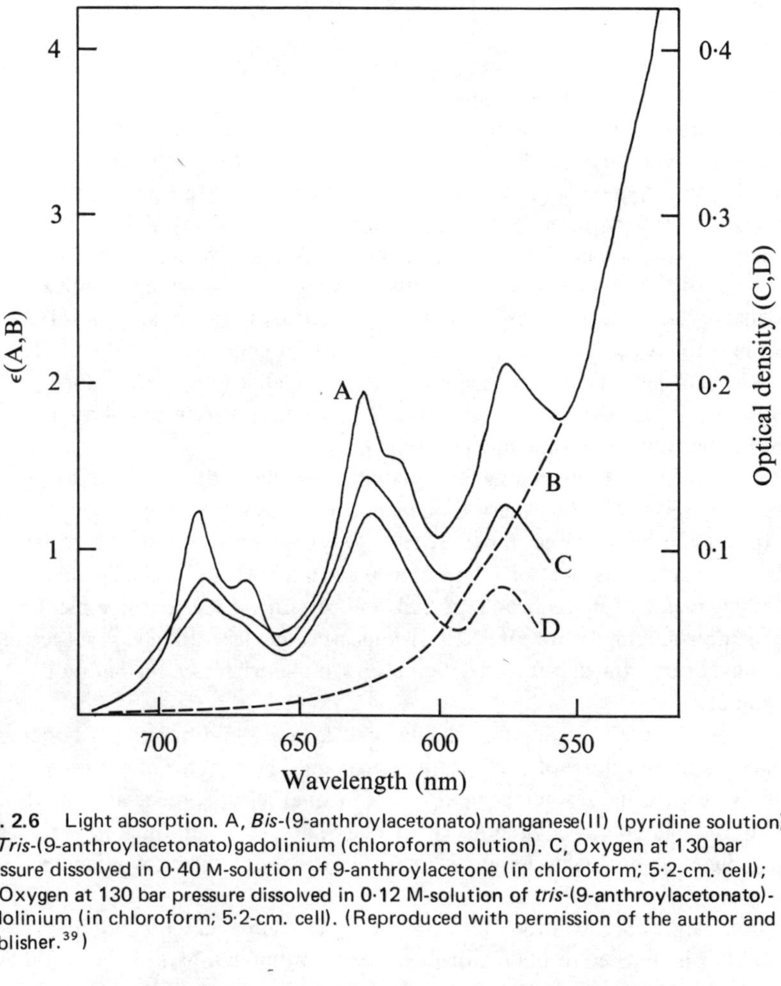


Fig. 2.6 Light absorption. A, *Bis*-(9-anthroylacetonato) manganese(II) (pyridine solution); B, *Tris*-(9-anthroylacetonato)gadolinium (chloroform solution). C, Oxygen at 130 bar pressure dissolved in 0·40 M-solution of 9-anthroylacetone (in chloroform; 5·2-cm. cell); D, Oxygen at 130 bar pressure dissolved in 0·12 M-solution of *tris*-(9-anthroylacetonato)-gadolinium (in chloroform; 5·2-cm. cell). (Reproduced with permission of the author and publisher.[39])

festation is an enhancement of the forbidden $T_1 \leftarrow S_0$ absorption (the reverse of phosphorescence) which occurs, for example, when 9-anthroylacetone is complexed with $Mn^{2+}$, (Fig. 2.6).[36] The presence of the heavy atom induces a higher degree of singlet–triplet mixing in the wave functions of the electronic states, mainly by a spin–orbit interaction, with a concomitant breakdown of the $\Delta S = 0$ selection rule.

The introduction of external heavy atoms, for example in the form of a solvent, co-solute, or matrix can induce similar effects. $\tau_P$ for benzene, which is 16 s in an argon matrix at 4·2 K, falls to 1 and 0·07 s in krypton and xenon matrices respectively, and the values of the ratio $\varphi_F/[\varphi_P + \varphi_{ISC}(T_1 \rightarrow S_0)]$ falls from 0·6 ($CH_4$ matrix) to *ca.* 0·05 (Ar) and zero (Kr and Xe)[37]. $T_1 \leftarrow S_0$ absorptions are promoted by ultilizing ethyl iodide as solvent and saturation of ethanolic solutions of various aromatics with xenon increases $\varphi_T$ at the expense of $\varphi_F$.[38]

The introduction of oxygen at pressures of 100 bar into solutions of aromatics, alkenes, and alkynes induces the $T_1 \leftarrow S_0$ absorption (Fig. 2.6), which occurs in the same spectral region as phosphorescence, the reverse process; (in addition bands attributable to charge-transfer complexes between the aromatic (donor) and $O_2$ (acceptor) can also be observed). In some cases identical absorptions have been obtained both by using ethyl iodide solvent as the perturbing agent and by introducing high pressures of paramagnetic $NO^{\bullet}$. The origin of the perturbation is not purely magnetic, however, as complexing of 9-anthroylacetone with the highly paramagnetic dysprosium ion ($\chi_P$ = 10·5 *B.M.*) fails to induce the forbidden transition, whilst complexing with Mn(II) ($\chi_P$ = 5·9 *B.M.*) and $O_2$ perturbation are highly effective (Fig. 2.6);[39] the importance of charge-transfer interaction between triplet state and perturbing agent has been introduced as an additional factor.

The problem of disentangling 'heavy atom' and 'magnetic' perturbations is particularly severe for the observed quenching of phosphorescence, $T_2 \leftarrow T_1$ absorption and photochemical reactivity of polyacenes and aromatic ketones by transition-metals ions. A strong dependence upon the nature of the ligand of the quenching metal atom has been reported, and certain *diamagnetic* complexes, e.g., bis(dipivaloylmethanato)nickel(II), appear to quench strongly,[40] suggesting collisional energy-transfer from triplet ketone to quencher as a further mode (section 3.1).

It is found quite generally that the phosphorescence lifetimes of perdeuterated polyacenes exceed those of their protium analogues by a factor of between 1·5 and 8 in a crystalline argon matrix at 4·2 K, in an ether-*iso*pentane-alcohol glass at 77 K and in a polymethylmethacrylate plastic at 298 K, although $\tau_P$ is somewhat reduced at the higher temperature in all cases, but at most by a factor of two.

The enhanced deactivation of $T_1$ for protio, as compared with deutero, compounds is ascribed to the relatively lower quantum numbers of the vibrational levels of the $S_0$ state isoenergetic with the $T_1(v = 0)$ level (Fig. 2.4). These vibrational levels will have relatively fewer nodes in their wave functions and the resulting greater overlap with the vibrational wave function of the $T_1(v = 0)$ state will increase the probability of inter-state crossing.[41]

The competition between fluorescence of a molecule A and all the other processes open to its $S_1$ state, both spontaneous and those brought about by an added substance Q, which may 'quench' by enhancing ISC or by chemical interaction, can be expressed in terms of the Stern–Volmer equation. Writing the competing processes,

$$^{1}\mathrm{A} \xrightarrow{h\nu} {}^{1}\mathrm{A}^*$$

$$^{1}\mathrm{A}^* \xrightarrow{k_F} {}^{1}\mathrm{A} + h\nu_F \qquad \text{(fluorescence)}$$

$$^{1}\mathrm{A}^* \xrightarrow{k_D} {}^{1}\mathrm{A} \quad \text{(radiationless deactivation)}$$

$$^{1}\mathrm{A}^* + \mathrm{Q} \xrightarrow{k_Q} {}^{1}\mathrm{A} + \mathrm{Q}\ (\text{or } \mathrm{Q}^*) \quad \text{(quenching)}$$

then

$$\varphi_F = \frac{\text{quanta emitted}}{\text{quanta absorbed}} = \frac{k_F[^1\mathrm{A}^*]}{k_F[^1\mathrm{A}^*] + k_D[^1\mathrm{A}^*] + k_Q[^1\mathrm{A}^*][\mathrm{Q}]}$$

The ratio of the fluorescence yields in the presence and absence of quencher, $\varphi_F$ and $\varphi_F{}^0$ respectively, gives the Stern–Volmer equation,

$$\frac{\varphi_{F0}}{\varphi_F} = 1 + \tau k_Q[\mathrm{Q}]$$

where $\tau = 1/(k_F + k_D)$. A simple graphical plot affords values for $\tau k_Q$ and direct measurement of $\tau$, the lifetime of $^1$A* in the absence of quencher, enables estimation of $k_Q$.

## 2.5 Excitation of complexes

The term 'complex' is used in its broadest sense, referring not only to donor-acceptor systems (existing in the ground state, in addition to an excited state) and to coordination compounds, involving strong covalent bonding between a central metal atom and surrounding ligands, but also to ions of uncertain solvation number, e.g., aqueous iodide ion. In all of these examples we are dealing with a complex species formed by chemical interaction of varying type in the ground state between simpler species capable of independent existence (and photochemistry). It is to be expected that the electrons participating in the bonding between the components will be featured strongly in the excited states of the complex. Molecular charge-transfer complexes provide a particularly simple illustration of this field and serve as an introduction to the more involved photochemistry of metal complexes.

### *2.5.1 Molecular charge-transfer (CT) complexes*

The bonding in these has been discussed by Mulliken in terms of ground-state and first singlet-state wave functions, $\psi(S_0)$ and $\psi(S_1)$, of the type,

$$\psi(S_0) = a\psi_0(\mathrm{D, A}) + b\psi_1(\mathrm{D}^+, \mathrm{A}^-) \qquad (2.14)$$

$$\psi(S_1) = a^*\psi_1(\mathrm{D}^+, \mathrm{A}^-) - b^*\psi_0(\mathrm{D, A}) \qquad (2.15)$$

where $\psi_0$ corresponds to a structure involving only weak physical forces such as dipole–dipole interaction and $\psi_1$ corresponds to a structure where one electron has been completely transferred from the donor (D) to the acceptor (A). $b/a$ and $b^*/a^*$ are normally both very small. The excitation $S_1 \leftarrow S_0$ is allowed and, in the case of both strong donor and strong acceptor, is found at longer wavelength than the absorptions of the components in accordance with the several theoretical treatments of the subject.[42] Characteristic charge-transfer fluorescence corresponding to the reverse transition, $S_1 \rightarrow S_0$, has been measured more frequently in low-temperature glasses but also in fluid solution at 300 K, and phosphorescence from the corresponding $T_1$ state has been detected in some instances.

The photochemical importance of the CT complexes is apparent from irradiation in the CT absorption band using filters, the simplest mode of reaction being ionization (to radical ions),

$$(\mathrm{D, A}) \xrightarrow{h\nu} \mathrm{D}^{\dot{+}} + \mathrm{A}^{\dot{-}} \tag{2.16}$$

This process has been characterized by the appearance of the e.s.r. spectrum of $\mathrm{A}^{\dot{-}}$ during photolysis of solutions of tetracyanoethylene (acceptor) in tetrahydrofuran (solvent donor).[43] The reaction products from photolysis of CT complexes of cyclohexene (donor) with various acceptors have been determined and an example is discussed in section 4.3.6.

## 2.5.2 *Solvated ions*

Solvation of ions in solution may involve either strong bonding between the ion and a specific number of solvent molecules, indicated by lack of exchange say, between, $^{18}$O-labelled hydrate and ordinary water, or weaker bonding to an unspecified number of rapidly exchanging solvent molecules. Even in the latter case, however, the absorption spectrum of the ion may be highly sensitive to the composition of the solvent and may be found at a wavelength different from that obtained in the gas phase. The solvent is considered to participate in the absorption process to give a charge-transfer transition in which the electron is accepted not into the orbitals of a single molecule but rather into a potential well defined by a group of several solvent molecules; accordingly these spectra are denoted ‘charge-transfer-to-solvent’ or CTTS spectra.[44] For example, the absorption of aqueous iodide ion occurs as a doublet at energies of 529·3 and 618·4 kJ mol$^{-1}$, corresponding closely to the energy difference in the gas phase of the $^2P_{1/2}$ and $^2P_{3/2}$ states of 90·92 kJ mol$^{-1}$ for iodide ion. The processes of ionization and deactivation are related as follows,

$$\mathrm{X}^-_{\mathrm{aq}} \xrightarrow{h\nu} (\mathrm{X}^-_{\mathrm{aq}})^*$$

$$(\mathrm{X}^-_{\mathrm{aq}})^* \rightarrow \mathrm{X}^-_{\mathrm{aq}} \qquad (\mathrm{X}^-_{\mathrm{aq}})^* \rightarrow (\mathrm{X}^{\bullet} + \mathrm{e}^-)_{\mathrm{aq}} \rightleftharpoons \mathrm{X}^{\bullet}_{\mathrm{aq}} + \mathrm{e}^-_{\mathrm{aq}}$$

$\varphi(\mathrm{e}^-_{\mathrm{aq}})$ measured by reaction with scavengers is about 0·3 and the characteristic spectrum of the solvated electron, $\mathrm{e}^-_{\mathrm{aq}}$ ($\lambda_{max} \sim 700$ nm), is obtained on flash photolysis of aqueous potassium iodide and chloride (Fig. 2.7)[45] along with spectra of the halogen radical-ions, $X_2^{\cdot-}$, formed by attack of $X^{\cdot}$ upon $X^-$.

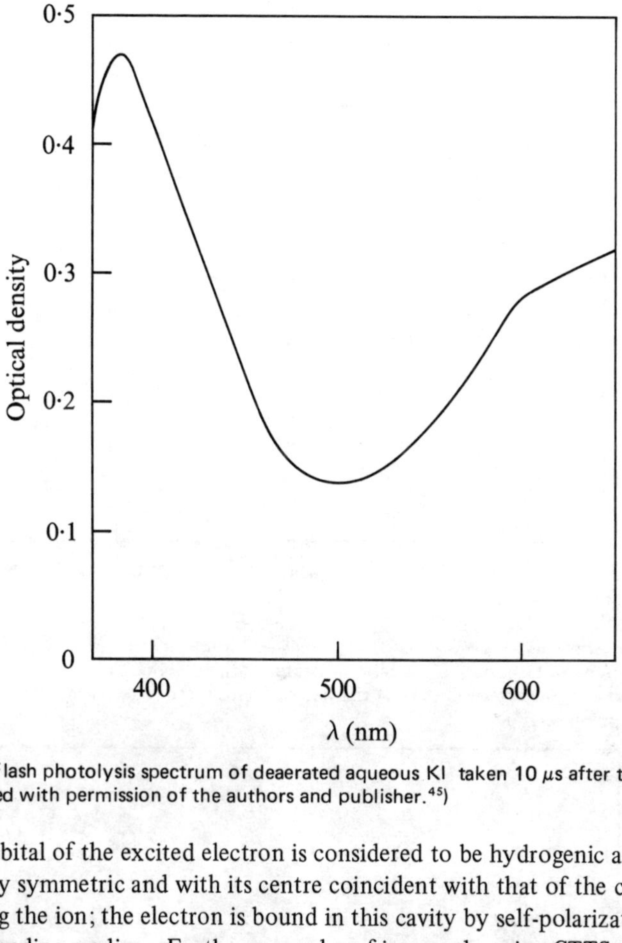


**Fig. 2.7** Flash photolysis spectrum of deaerated aqueous KI taken 10 μs after the flash. (Reproduced with permission of the authors and publisher.[45])

The orbital of the excited electron is considered to be hydrogenic and spherically symmetric and with its centre coincident with that of the cavity containing the ion; the electron is bound in this cavity by self-polarization of the surrounding medium. Further examples of ions undergoing CTTS processes are given in Table 2.6.[46–50]

Whilst photolysis of *reducing* ions has been shown to result in transfer of a solute electron to the solvation shell, that of *oxidizing* ions might be expected to produce the reverse of this and flash photolysis of aqueous ceric nitrate and sulphate suggests the following process[51]

$$Ce(IV)_{aq} \xrightarrow{h\nu} Ce(III) + OH^{\cdot} + H^+ \qquad (2.17)$$

followed by

$$OH^{\cdot} + HSO_4^- \rightarrow OH^- + HSO_4^{\cdot} \qquad (2.18)$$

$$OH^{\cdot} + NO_3^- \rightarrow OH^- + NO_3^{\cdot} \qquad (2.19)$$

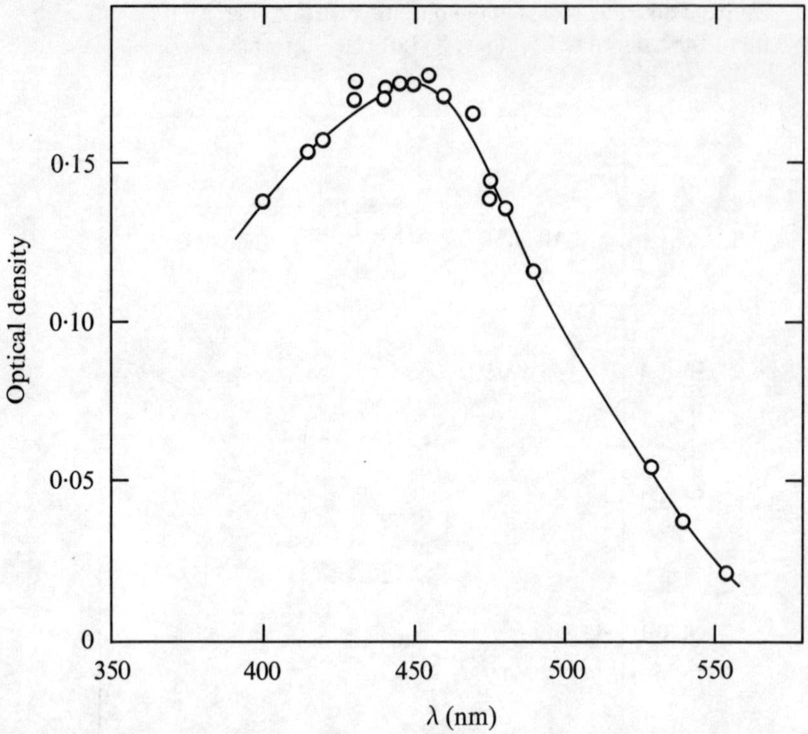


**Fig. 2.8 (a)** Absorption spectrum of $HSO_4^{\bullet}$ transient produced on flash photolysis of aerated aqueous solutions of $10^{-4}$ M ceric sulphate in 1·0 M $H_2SO_4$. The optical density was measured at 40 $\mu$s after the start of the flash. (Reproduced with permission of the authors and publisher.[51])

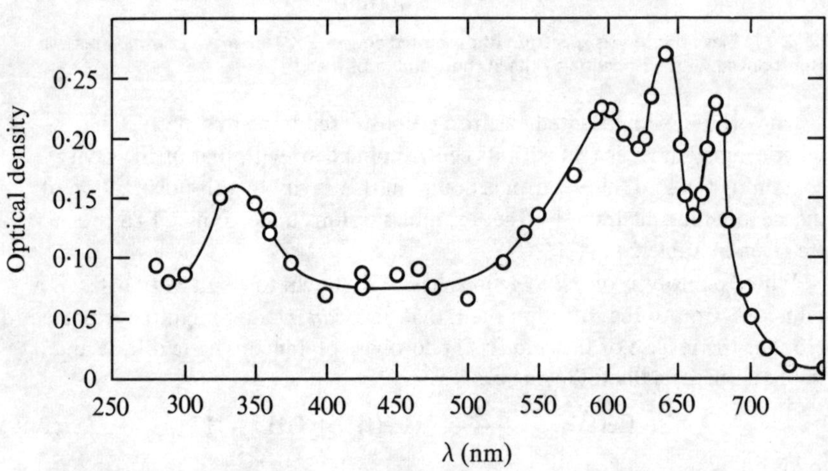


**Fig. 2.8 (b)** Absorption spectrum of $NO_3^{\bullet}$ transient produced on flash photolysis of aerated aqueous solutions of 0·1 M potassium ceric nitrate. The optical density was measured at 200 $\mu$s after the start of the flash. (Reproduced with permission of the authors and publisher.[51])

The absorption spectra of the solutions, attributed to the inorganic radicals, are shown in Fig. 2.8. Process (2.17) is favoured over charge-transfer from sulphate and nitrate ligands because of the depressant effect of $OH^{\bullet}$ scavengers upon the absorptions at 'zero time,' but it can still be regarded as an example of a charge-transfer-to-metal (CTTM) process (section 2.5.3).

Flash photolysis of aqueous solutions of halate ions produces spectra assigned to $XO_2^{\bullet}$ species (X = Cl, Br, I), suggesting a process[52]

$$XO^-_{3\,aq} \xrightarrow{h\nu} (XO^-_{3\,aq})^* \rightarrow XO_2^{\bullet} + OH^{\bullet} + OH^-$$

Flash photolysis of $NO_3^-$, however, produces peroxynitrite ion, $O{-}N{-}O{-}O^-$ which grows in after the flash; the nature of the primary act is obscure.[53] Peroxydisulphate ion, $S_2O_8^{2-}$, provides a further contrast in undergoing homolytic fission,[54] cf. section 2.3,

$$S_2O^{2-}_{8\,aq} \xrightarrow{h\nu} 2SO_4^{\bar{\cdot}}$$

TABLE 2.6: CTTS processes in aqueous solution

| Ion | λ (nm) | Process | φ | Technique | Reference |
|---|---|---|---|---|---|
| $X^-$ | 184·9 | $X^{\bullet} + e_{aq}^-$ | 0·43 (X = Cl) | Scavenging of $e_{aq}^-$ by $N_2O$ | 46 |
| (X = Cl, Br) | | | 0·34 (X = Br) | Flash photolysis | 45 |
| $I^-$ | 253·7 | $I^{\bullet} + e_{aq}^-$ | 0·32 | Scavenging of $e_{aq}^-$ by $N_2O$ | 47 |
| | | | | Flash photolysis | 45 |
| $CNS^-$ | | $CNS^{\bullet} + e_{aq}^-$ (competing with $CN^- + S$) | | Flash photolysis | 48 |
| $N_3^-$ | | $N_3^{\bullet} + e_{aq}^-$ | | Flash photolysis | 49 |
| $OH^-$ | 184·9 | $OH^{\bullet} + e_{aq}^-$ | 0·11 | Scavenging of $e_{aq}^-$ by $N_2O$ | 46 |
| | | | | Flash photolysis | 50 |
| $SO_4^{2-}$ | 184·9 | $SO_4^{\bar{\cdot}} + e_{aq}^-$ | 0·71 | Scavenging of $e_{aq}^-$ by $N_2O$ | 46 |
| | | | | Flash photolysis | 50 |
| $SO_3^{2-}$ | | $SO_3^{\bar{\cdot}} + e_{aq}^-$ | | Flash photolysis | 48 |
| $S_2O_3^{2-}$ | | $S_2O_3^{\bar{\cdot}} + e_{aq}^-$ (competing with $S_2O_2^{\bar{\cdot}} + OH^{\bullet} + OH^-$) | | Flash photolysis | 48 |
| $CO_3^{2-}$ | | $CO_3^{\bar{\cdot}} + e_{aq}^-$ | | Flash photolysis | 50 |

(See also Table 2.7)

### *2.5.3 Metal complexes*

The photochemistry of certain transition metal ions has been studied in sufficient depth to permit some generalization. The spectra of complexes of these ions feature two principal types of transition namely a set of long wave-

TABLE 2.7: Examples of the photochemistry of transition metal complexes

| Substrate | λ(excitation) (nm) | Product | φ | Reference |
|---|---|---|---|---|
| *trans*-$[Cr(NH_3)_2(NCS)_4]^-$ | 316–750 | $NCS^-$ | ~ 0·3 | 55 |
| $Cr(NH_3)_6^{3+}$ | 320–600 | $[Cr(NH_3)_5(H_2O)]^{3+}$ | 0·32 | 56 |
| | 650 | | ~ 1 | 57 |
| $Cr(C_2O_4)_3^{3-}$ | 420–697 | racemate | 0·07–0·09 | 58 |
| $M(CO)_6$ (M = Cr, W, Mo) | | | | |
| (a) in hydrocarbon matrix at 77 K | 366 | $M(CO)_5$ | 1·0 | 59 |
| (b) in presence N-base (L) (L = piperidine, triethylamine) | 366 | $M(CO)_5L$ | 1·0 | 60 |
| $Mo(CN)_8^{4-}$ | 366 | $[Mo(CN)_7(H_2O)]^{3-} + CN^-$ | 0·14 | 61 |
| $M(CN)_8^{4-}$ (M = Mo, W) | Kr flash | $M(CN)_8^{3-} + e_{aq}^-$ | – | 62 |
| $Mn(C_2O_4)_3^{3-}$ | 313–400 | $Mn(C_2O_4)_2^{2-}$ | 0·50 | 63 |
| $MnO_4^-$ | 253·7 | $MnO_2 + O_2^-$ | 0·054 | 64 |
| | 366 | or | 0·0021 | |
| | 564 | $MnO_2^- + O_2$ | $1·4 \times 10^{-5}$ | |

| | | | | |
|---|---|---|---|---|
| $Fe(CN)_6^{4-}$ | 366 | $[Fe(CN)_5(H_2O)]^{3-} + CN^-$ | 0·52(pH 0·65)<br>0·89(pH 4·0) | 65 |
| | 253·7 | $Fe(CN)_6^{4-} + e_{aq}^-$ | 0·66 | 66 |
| $Fe_{aq}^{2+}$ | 253·7 | $Fe_{aq}^{3+} + e_{aq}^-$ | 0·07 | 66 |
| $Fe(CO)_5$ in presence N-base (L) (L = pyridine, picoline, piperidine) | u.v. | $Fe(CO)_4L$ (L *axial*) | – | 67 |
| $(C_5H_5)_2Fe$ (Solvent $CCl_4$) | CT region (~ 310) | $(C_5H_5)_2Fe^+$ | – | 68 |
| $[Co(NH_3)_5I]^{2+}$ | 340–400<br>550 | Co(II) + 1/2 $I_2$ | 0·66<br>0·10 | 69 |
| $[Co(NH_3)_5Cl]^{2+}$ | 340–400 | *trans*-$[Co(NH_3)_4(H_2O)(Cl)]^{2+}$ | 0·011 | 69, 70 |
| *cis*-$[Pt(NH_3)_2Cl_2]$ (298 K) | 350<br>363<br>410 | *cis*-$[Pt(NH_3)_2(H_2O)Cl]^+$ | 0·46<br>0·39<br>0·25 | 71 |
| *cis*- and *trans*-$[Pt(PEt_3)_2Cl_2]$ | > 304 | *cis-trans* mixture | 0·01–0·02 | 72 |
| $PtCl_6^{2-}$ | wide range | $[PtCl_5(OH)]^{2-}$ | – | 73 |
| | various (230–546) | Exchange with $^{36}Cl^-$ and $Br^-$ | 15 to 1312 | 74 |
| $PtBr_6^{2-}$ | 313, 365, 433, and 530 | $[PtBr_5(H_2O)]^-$ | 0·4 | 75 |
| | 540 | Exchange with $^{80}Br^-$ | 10–500 (dependent on $I_a$) | 76 |

length (> 300 nm) absorptions (sometimes structured) of low extinction and a structureless absorption of high extinction at short wavelength (< 250 nm). The former group, which are forbidden, are designated *d-d* transitions in crystal-field terminology, or as ligand-field (LF) transitions in ligand field theory. The u.v. transition is parity-allowed and is denoted a charge-transfer transition because it involves movement of electron density either towards the metal M, (to give a charge-transfer-to-metal or CTTM absorption), or towards the ligand L (a CTTL absorption) or, finally, to the solvent molecules (a CTTS absorption as in the cases of some simple anions). Both of the principal types of transition are significant in the photochemistry of these compounds, although, as one might expect, the chemical processes induced by irradiating in the shorter wavelength CT band are normally of higher energy than those obtained with the LF bands, CT excitation leading to net movement of an electron to give oxidized or reduced metal plus a solvent or ligand radical fragment, but LF excitation resulting in a heterolytic process, a familiar example being solvolysis of one ligand. Other reactions appear to depend on radiationless conversion from excited states to give a vibrationally excited ground state which can undergo the normal slow 'thermal' reaction of the complex with greater facility. The examples of Table 2.7 illustrate these various processes.[55–76]

The correlation of redox and substitution reactions with u.v. and visible absorptions respectively is imperfect, however, and filtered irradiation of the *d-d* band of trisoxalato complexes of Mn(III) and Fe(III) produces efficient redox breakdown[63] (equation (2.20)) instead of the racemization found with the analogous Cr(III) complex,[58]

$$2\,M(C_2O_4)_3{}^{3-} \xrightarrow{h\nu} 2\,M(II) + 5\,C_2O_4{}^{2-} + 2\,CO_2 \qquad (2.20)$$

One possible explanation is that the *d-d* bands contain sufficient admixed CT character that the CT process (2.21) can occur in the *d-d* absorption region[63]

$$M^{III}(C_2O_4)_3{}^{3-} \xrightarrow[\lambda \sim 500\ nm]{h\nu} [M^{II}(C_2O_4)_2(C_2O_4{}^{\bar{\cdot}})]^{3-}$$

$$\downarrow$$

$$M(C_2O_4)_2{}^{2-} + C_2O_4{}^{\bar{\cdot}} \qquad (2.21)$$

Within a given series of complexes differing only by a single ligand, for example $[Co(NH_3)_5X]^{2+}$ where $X = F^-, Cl^-, Br^-, I^-, N_3{}^-, NO_2{}^-, SCN^-$, an additional correlation is found between $\varphi_{redox}$ at a given wavelength and the ease of thermal oxidation of X; moreover $\varphi_{aq}$, $\varphi_{redox}$, and $\varphi_{redox}/\varphi_{aq}$ decrease as the wavelength of irradiation is increased from 370 nm (nominal) to 600 nm (Table 2.8).[69] The primary act proposed for the redox process,

$$[Co(NH_3)_5X]^{2+} \xrightarrow{h\nu} Co(II) + 5NH_3 + X^{\cdot}$$

is supported by the observation of halogen atom transients in flash photolysis of aqueous solutions (X = Br, I).[77] The co-existence of aquation and redox processes evident, for example, in the photolysis of the bromo-complex (Table 2.8) cannot readily be dismissed as mere mixing of CT and LF transitions for

TABLE 2.8: Photochemistry of the ions $[Co(NH_3)_5X]^{n+}$ (298 K, pH 4)[69]

| X | 370 nm irradiation $\varphi_{redox}$ | 370 nm irradiation $\varphi_{aq}$ | 550 nm irradiation $\varphi_{redox}$ | 550 nm irradiation $\varphi_{aq}$ |
|---|---|---|---|---|
| $F^-$ | small | small | small | small |
| $Cl^-$ | trace | 0·011 | trace | $1{\cdot}5 \times 10^{-3}$ |
| $Br^-$ | 0·105 | 0·105 | trace | $1{\cdot}3 \times 10^{-3}$ |
| $NO_2^-$ | 0·65 | 0·35 | – | – |
| $SCN^-$ | 0·0315 | 0·0135 | $1{\cdot}3 \times 10^{-4}$ | $5{\cdot}4 \times 10^{-4}$ |
| $N_3^-$ | 0·44 | $< 4 \times 10^{-3}$ | 0·011 | $< 10^{-4}$ |
| $I^-$ | 0·66 | $< 6 \times 10^{-3}$ | 0·10 | $< 10^{-3}$ |

irradiation in the LF bands produces a much lower $\varphi_{aq}$ than for Cr(III) complexes. The aquation may possibly result from failure of $X^{\bullet}$ to escape the solvent cage following CT excitation;[78]

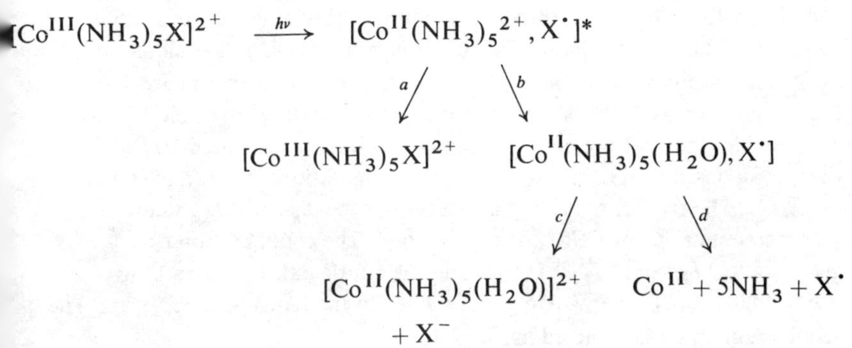


$X^{\bullet}$ is immediately recaptured (step *a*) or begins to diffuse away (step *b*), the caged radical-pair then diffuse apart (step *d*) or exchange an electron without return of X into the coordination sphere (step *c*). Decrease of irradiation wavelength increases the degree of excitation of the initial product, promoting the rapid separation of the radical-pair, i.e., to increase both $\varphi_{redox}$ and $\varphi_{redox}/\varphi_{aq}$.

In contrast to the largely redox behaviour of complexes of Co(III) on photolysis, those of Cr(III) show almost exclusively substitution reactions, a differentiation linked to the relative thermal stabilities of these complexes towards redox breakdown and to the shorter wavelength of the CT bands of Cr(III) complexes.

Pt(II) complexes undergo largely substitution or isomerization reactions irrespective of the bands irradiated (LF or CT); some examples are given in Table 2.7. $Pt^{IV}Cl_6{}^{2-}$ in addition to undergoing photoaquation to $[PtCl_5(OH)]^{2-}$ also exchanges with $^{36}Cl^-$ and $Br^-$ with $\varphi$ of 15 to $> 10^3$, depending on wavelength, pH, and temperature. $PtBr_6^{2-}$ shows entirely similar behaviour (Table 2.7) and a general chain mechanism for the exchange process can be formulated,

$$PtX_6^{2-} \xrightarrow{h\nu} PtX_5^{2-} + X^\bullet$$

$$PtX_5^{2-} + {}^*X^- \longrightarrow PtX_4^*X^{2-} + X^-$$

$$PtX_4^*X^{2-} + PtX_6^{2-} \longrightarrow PtX_5^*X^- + PtX_5^{2-}$$

Complexes of non-transition metals are free of LF effects and their photochemistry is dominated by CT processes, for example, flash photolysis of halides of Hg(II) produces[79] the transient $X_2^{\bar{\bullet}}$

$$HgX_2 \xrightarrow{h\nu} [Hg^{\dot{+}} X_2^{\bar{\bullet}}] \rightarrow Hg(I) + X_2^{\bar{\bullet}}$$

Whilst it is possible to give a qualitative discussion of the factors leading to redox or substitution processes, quantitative descriptions of the excited states of metal complexes and the relationship of these with photochemical action is rendered difficult by paucity of data on the MO's involved. For a complex such as $Co(NH_3)_6^{3+}$ the metal–ligand bonding is sigma in character and a set of MO's can be drawn up as in Fig. 1.7. The $3d_{z^2}$, $3d_{x^2-y^2}$, $4s$, and $4p$ orbitals combine with six ligand $\sigma$ orbitals to produce six bonding and six antibonding MO's. Should ligand $\pi$ orbitals also be present then these will combine with the $t_{2g}$ metal $d$ orbitals ($3d_{xy}$, $3d_{xz}$, and $3d_{yz}$) to produce further MO's. The sequence of energy levels is determined by the strength of the ligand field, measured by the splitting of the $e_g$ and $t_{2g}$ orbitals and denoted $10\,Dq$. For $CrL_6^{n-}$, an octahedral complex with a $d^3$ configuration, the ground state is ${}^4A_{2g}, t_{2g}^3$. The $t_{2g}^3$ configuration also gives rise to ${}^2E_g$ and ${}^2T_{1g}$ states (degenerate in a strong field) and a ${}^2T_{2g}$ state. The configuration $t_{2g}^2\ e_g^1$ gives rise to ${}^4T_{2g}$ and ${}^4T_{1g}(F)$ states and the configuration $t_{2g}^1\ e_g^2$ gives rise to the ${}^4T_{1g}(P)$ state, where $F$ and $P$ refer to the terms of the free ion. The sequence of states is depicted in Fig. 2.9.

Aqueous $Cr(H_2O)_6^{3+}$ shows weak absorptions ($\epsilon \sim 15$) at 575, 408, and 263 n[m] assigned to the transitions ${}^4T_{2g} \leftarrow {}^4A_{2g}$, ${}^4T_{1g}(F) \leftarrow {}^4A_{2g}$, and ${}^4T_{1g} \leftarrow {}^4A_{2g}$ respe[ct]ively, which are spin-allowed but parity-forbidden. The spin- and parity-forbidden low-energy transitions to the various doublet states ($\epsilon \sim 1$) are found in the 670 nm region, but are often buried in the tails of other bands. Luminescence yields from Cr(III) complexes are low even in glassy solution at 77 K; $\varphi_F$ is zero except in a few cases, for example urea- and aquo-complexes, for which $\varphi_F = 0{\cdot}00213$ and *ca.* $10^{-4}$ respectively, and $\varphi_P$ is usually less than 0·01, but reaches 0·23 in $[Cr(NCS)_6]^{3-}$ and 1 in ruby. The fluorescence is assigned to a ${}^4T_{2g} \rightarrow {}^4A_{2g}$ transition and the phosphorescence, which is independent of excitatio[n] wavelength, is assigned to a ${}^2E_g \rightarrow {}^4A_{2g}$ transition (Fig. 2.9). Very fast radiationless transitions from the ${}^2E_g$ state account for the low $\varphi_P$.

The thermally equilibrated ${}^4T_{2g}$ and ${}^2E_g$ states are those most likely to be the reactive intermediates in solvolysis following LF excitation because of the independence of $\varphi_{aq}$ upon excitation wavelength. Selective excitation of the quartet and doublet states in the cases of *trans*-$[Cr(NH_3)_2(NCS)_4]^-$ produces no change in $\varphi_{aq}$ (Table 2.7). This implies rapid inter-system crossing to the lower state which is photoreactive, which is in most cases the $^2E_g$ state. However, this point remains controversial and both $^4T_{2g}$ and $(^4A_{2g})^v$ states have been proclaimed photoreactive.

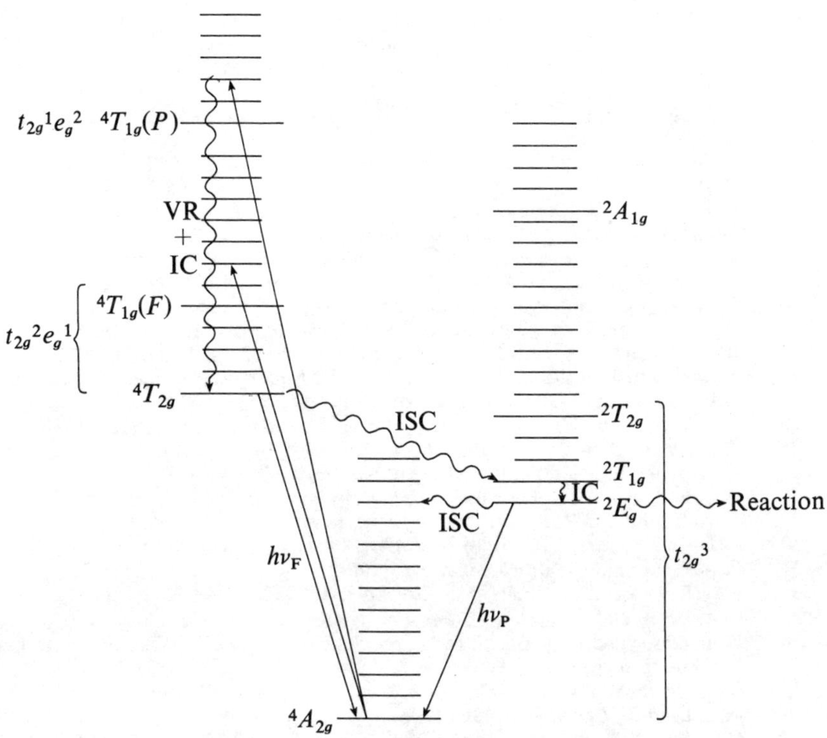


**Fig. 2.9** Jablonski diagram for lower states of octahedral Cr(III) complex in strong field.

A similar state diagram can be drawn for octahedrally coordinated Co(III) which has a ground state $t_{2g}^6$ configuration giving a $^1A_{1g}$ term and a first-excited $t_{2g}^5\,e_g^1$ state giving $^3T_{1g}$, $^3T_{2g}$, $^1T_{1g}$, and $^1T_{2g}$ terms energy increasing in that order. The LF bands of Co(III) complexes at 500 and 350 nm respectively are $^1T_{1g} \leftarrow {}^1A_{1g}$ and $^1T_{2g} \leftarrow {}^1A_{1g}$ respectively. The doubly-forbidden triplet transitions are very weak. Luminescence from Co(III) complexes is found very rarely, possibly because of the change of electron configuration. As stated before, the photochemistry of Co(III) is dominated by CT processes.

## References


1. A. B. Callear and R. G. W. Norrish, *Proc. Roy. Soc.*, **A266**, 299 (1962).
2. P. H. Garrett, *Phys. Rev.*, **40**, 779 (1932).
3. M. L. Pool, *Phys. Rev.*, **35**, 1419 (1930).
4. E. W. R. Steacie, *Atomic and Free Radical Reactions*, 2nd edn., Chapter V, Reinhold (1954).
5. G. Cario and J. Franck, *Z. Phys.*, **17**, 202 (1923).
6. J. G. Calvert and J. N. Pitts, Jr., *Photochemistry*, Wiley, Chapter 2 (1966).
7. H. E. Gunning and O. P. Strausz, *Adv. Photochem.*, **1**, 209 (1963).
8. A. B. Callear and R. J. Cvetanović, *J. Chem. Phys.*, **24**, 873 (1956).
9. L. Mathieson and A. L. G. Rees, *J. Chem. Phys.*, **25**, 753 (1956).
10. E. J. Bowen, *Chemical Aspects of Light*, 2nd edn., Clarendon Press, Oxford (1946).
11. L. E. Compton, J. L. Gole, and R. M. Martin, *J. Phys. Chem.*, **73**, 1158 (1969).
12. H. W. Ford and S. Jaffe, *J. Chem. Phys.*, **38**, 2935 (1963).
13. R. P. Wayne, *Nature* (*London*), **203**, 516 (1964).
14. F. J. Lipscomb, R. G. W. Norrish, and B. A. Thrush, *Proc. Roy. Soc.*, **A233**, 455 (1956).
15. N. Basco and R. G. W. Norrish, *Proc. Roy. Soc.*, **A260**, 293 (1961).
16. G. Herzberg and D. A. Ramsay, *Disc. Faraday Soc.*, **14**, 11 (1953).
17. F. J. Adrian, E. L. Cochran, and V. A. Bowers, Chapter 5 in *Free Radicals in Inorganic Chemistry*, A.C.S. Monograph No. 36 (1962).
18. D. A. Ramsay, *J. Phys. Chem.*, **57**, 415 (1953).
19. R. G. W. Norrish and G. A. Oldershaw, *Proc. Roy. Soc.*, **A262**, 1 (1961).
20. W. A. Noyes, Jr. and P. A. Leighton, *The Photochemistry of Gases*, Reinhold (1941).
21. R P. Porter and W. A. Noyes, Jr., *J. Amer. Chem. Soc.*, **81**, 2307 (1959).
22. R. Gomer and W. A. Noyes, Jr., *J. Amer. Chem. Soc.*, **72**, 101 (1950).
23. Y. Takezaki, T. Miyazaki, and N. Nakahara, *J. Chem. Phys.*, **25**, 536 (1956).
24. J. T. Martin and R. G. W. Norrish, *Proc. Roy. Soc.*, **A220**, 322 (1953).
25. J. V. Michael and W. A. Noyes, Jr., *J. Amer. Chem. Soc.*, **85**, 1228 (1963).
26. T. Inaba and B. de B. Darwent, *J. Phys. Chem.*, **64**, 1431 (1960).
27. W. West and L. Schlessinger, *J. Amer. Chem. Soc.*, **60**, 961 (1938).
28. P. G. Barker and J. H. Purnell, *Trans. Faraday Soc.*, **66**, 163 (1970).
29. A. A. Gordus and R. B. Bernstein, *J. Chem. Phys.*, **30**, 973 (1959).
30. K. D. Cadogan and A. C. Albrecht, *J. Phys. Chem.*, **72**, 929 (1968).
31. C. A. Parker, *Adv. Photochem.*, **2**, 305 (1964).
32. P. G. Bowers and G. Porter, *Proc. Roy. Soc.*, **A299**, 348 (1967).
33. C. A. Parker and T. A. Joyce, *Trans. Faraday Soc.*, **65**, 2823 (1969).
34. F. Wilkinson and A. R. Horrocks, in *Luminescence in Chemistry*, ed. E. J. Bowen, Van Nostrand chapter 7, (1968).
35. D. R. Kearns and W. A. Case, *J. Amer. Chem. Soc.*, **88**, 5087 (1966).
36. D. F. Evans, *J. Chem. Soc.*, 1987 (1961).
37. M. R. Wright, R. P. Frosch, and G. W. Robinson, *J. Chem. Phys.*, **33**, 934 (1960).
38. A. R. Horrocks, A. Kearvell, K. Tickle, and F. Wilkinson, *Trans. Faraday Soc.*, **62**, 3393 (1966).
39. D. F. Evans, *Proc. Roy. Soc.*, **A255**, 55 (1960).
40. G. S. Hammond and R. P. Foss, *J. Phys. Chem.*, **68**, 3739 (1964).
41. G. W. Robinson and R. P. Frosch, *J. Chem. Phys.*, **37**, 1962 (1962).
42. R. Foster, *Organic Charge-Transfer Complexes*, Academic Press (1969).
43. D. Ilten and M. Calvin, *J. Chem. Phys.*, **42**, 3760 (1965).
44. M. J. Blandamer and M. F. Fox, *Chem. Revs.*, **70**, 59 (1970).
45. M. S. Matheson, W. A. Mulac, and J. Rabani, *J. Phys. Chem.*, **67**, 2613 (1963).
46. F. S. Dainton and P. Fowles, *Proc. Roy. Soc.*, **A287**, 312 (1965).
47. F. S. Dainton and S. R. Logan, *Proc. Roy. Soc.*, **A287**, 287 (1965).
48. L. Dogliotti and E. Hayon, *J. Phys. Chem.*, **72**, 1800 (1968).
49. F. Barat, B. Hickel, and J. Sutton, *Chem. Comm.*, 125 (1969).
50. E. Hayon and J. J. McGarvey, *J. Phys. Chem.*, **71**, 1472 (1967).
51. L. Dogliotti and E. Hayon, *J. Phys. Chem.*, **71**, 3802 (1967).

52. F. Barat, L. Gilles, B. Hickel, and J. Sutton, *Chem. Comm.*, 1485 (1969).
53. F. Barat, B. Hickel, and J. Sutton, *Chem. Comm.*, 125 (1969).
54. L. Dogliotti and E. Hayon, *J. Phys. Chem.*, **71**, 2511 (1967).
55. E. E. Wegner and A. W. Adamson, *J. Amer. Chem. Soc.*, **88**, 394 (1966).
56. M. R. Edelson and R. A. Plane, *J. Phys. Chem.*, **63**, 327 (1959).
57. M. R. Edelson and R. A. Plane, *Inorg. Chem.*, **3**, 231 (1964).
58. S. T. Spees and A. W. Adamson, *Inorg. Chem.*, **1**, 531 (1962).
59. I. W. Stolz, G. R. Dobson, and R. K. Sheline, *J. Amer. Chem. Soc.*, **85**, 1013 (1963).
60. W. Strohmeier and D. Von Hobe, *Z. Phys. Chem.* (*Frankfurt*), **34**, 393 (1962).
61. A. W. Adamson and J. R. Perumareddi, *Inorg. Chem.*, **4**, 247 (1965).
62. W. L. Waltz, A. W. Adamson, and P. D. Fleischauer, *J. Amer. Chem. Soc.*, **89**, 3923 (1967).
63. G. B. Porter, J. G. W. Doering, and S. Karanka, *J. Amer. Chem. Soc.*, **84**, 4027 (1962).
64. G. Zimmerman, *J. Chem. Phys.*, **23**, 825 (1955).
65. S. Ohno, *Bull. Chem. Soc. Japan*, **40**, 1765 (1967).
66. P. L. Airey and F. S. Dainton, *Proc. Roy. Soc.*, **A291**, 340 (1966).
67. E. H. Schubert and R. K. Sheline, *Inorg. Chem.*, **5**, 1071 (1966).
68. E. Koerner v. Gustorf, and F. -W. Grevels, *Fortsch. der Chem. Forsch.*, **13**, 366 (1969).
69. A. W. Adamson, *Disc. Faraday Soc.*, **29**, 163 (1960).
70. L. Moggi, N. Sabbatini, and V. Balzani, *Gazz. Chim. Ital.*, **97**, 980 (1967).
71. J. R. Perumareddi and A. W. Adamson, *J. Phys. Chem.*, **72**, 414 (1968).
72. P. Haaki and T. A. Hylton, *J. Amer. Chem. Soc.*, **84**, 3774 (1962).
73. E. H. Archibald, *J. Chem. Soc.*, 1104 (1920).
74. R. Dreyer, K. König, and H. Schmidt, *Z. Phys. Chem.* (*Leipzig*), **227**, 257 (1964).
75. V. Balzani, F. Manfoin, and L. Moggi, *Inorg. Chem.*, **6**, 354 (1967).
76. A. W. Adamson and A. H. Sporer, *J. Amer. Chem. Soc.*, **80**, 3865 (1958).
77. S. A. Penkett and A. W. Adamson, *J. Amer. Chem. Soc.*, **87**, 2514 (1965).
78. A. W. Adamson and A. H. Sporer, *J. Inorg. Nucl. Chem.*, **8**, 209 (1958).
79. M. E. Langmuir and E. Hayon, *J. Phys. Chem.*, **71**, 3808 (1967).

# 3. Reactions of species produced photochemically—I

The simplest *reactions* undergone by small molecules following photoexcitation can be classified (*a*) as photophysical, i.e., involving the transfer of energy to some second molecule, possibly of the same type, (*b*), as radical, i.e., those depending either on the production of small free radicals familiar as intermediate in thermal, and especially chain reactions, in the gas phase and in solution, or on abstraction of an atom from another molecule, normally the solvent, and (*c*) as non-radical, e.g., isomerization, photoaddition, and elimination reactions. A particularly simple example of radical production is that of photoionization to produce an electron and a radical-cation

$$M \xrightarrow{h\nu} M^{\dot{+}} + e^-$$

The extrusion of small molecules such as $N_2$ is dealt with in chapter 4 and the atom-sensitized reactions, particularly of mercury, have been discussed in section 2.1.

## 3.1 Energy transfer

The processes

$$A^* + A \xrightarrow{k_{A^*A}} A^* + A \quad (3.1)$$

$$A^* + B \xrightarrow{k_{AB}} A + B^* \quad (3.2)$$

where A is an excited molecule of unspecified multiplicity, can involve the emission of a quantum of radiation from A* followed by reabsorption by A or B. This is denoted the 'trivial' process and, although it has been proposed as a possibly more universal means of energy transfer in condensed systems, this is no longer generally accepted and we shall not discuss it further except to note that it is the natural means of transmitting radiant energy between remote bodies.

An alternative physical picture for processes (3.1) and (3.2) is that of non-radiative transfer following collision between A* and A (or B). This process vould be expected to have a rate limited by the collision frequency of the two pecies, $Z_{AB}$. Such a limitation has been found in both gas-phase and solution tudies of the sensitized phosphorescence of biacetyl. Ishikawa and Noyes[1] xamined the 253·7 nm irradiated benzene (donor)–biacetyl (acceptor) system in he gas phase. At this wavelength all biacetyl emission is due to transfer from riplet excited benzene molecules, singlet excited benzene producing initially a ibrationally excited $S_2$ state of biacetyl which either dissociates to products, ndergoes internal conversion to $S_1$ and $S_0$, or crosses to the $T_2$ state and hence to $S_0$. The observed variation of $\varphi_P$ with biacetyl pressure at a constant enzene pressure fits very well with the kinetic scheme proposed. Mercury-hotosensitized reactions also depend on collision although the efficiency of eaction following collision is often much less than unity.

Singlet–singlet transfer, e.g., from naphthalene, to give sensitized biacetyl luorescence in solution has been measured by Dubois and colleagues[2] and heir data, taken with a measured value for the excited donor singlet lifetime, roduces transfer rates of the order of $3{\cdot}5 \times 10^{10}$ l mol$^{-1}$ s$^{-1}$ for hexane olutions at 301 K. These are close to those predicted by the modified Debye quation

$$k = \frac{8RT}{2000\eta} \tag{3.3}$$

nd transfer may be assumed to be 100% effective following collision.

Yet another mode of energy transfer corresponding to reactions 3.1 and 3.2 s that of non-collisional or long-range non-radiative transfer. This is of great mportance in understanding a number of general phenomena including the uminescent properties of molecular solids and the transfer of energy to solute pecies following radiolysis (radiation-chemical action induced by $\alpha$ and $\beta$ articles or $\gamma$ rays). The effectiveness of this process depends on several factors ncluding the spin multiplicity of the donor and acceptor, the extent of eparation of the two energy levels involved and the degree of order in the ntervening phase.

Long-range transfer of singlet energy in fluid solution between like solute nolecules was proposed by Perrin[3] to account for the decrease in the degree of olarization of fluorescent light at concentrations $> 10^{-3}$ M when fluorescent lye solutions are irradiated with polarized light. The onset of depolarization at his concentration corresponds to transfer of singlet energy from the donor over n average intermolecular separation of solute molecules of $\sim 7$ nm to give an xcited acceptor having no 'memory' of the direction of polarization of the ncident light beam. Long-range transfer between unlike molecules in both luid and glassy (disordered solid) environments was demonstrated by Bowen nd colleagues[4,5] using weakly fluorescent 1-chloroanthracene as donor $\varphi_F \sim 0{\cdot}075$) and strongly fluorescent perylene as acceptor. The strong uorescence from perylene in the mixed solution at low acceptor concentrations rovides strong evidence for efficient singlet–singlet transfer and the concentration dependence studies (Fig. 3.1) together with data for $\varphi_F$ and $\tau$ for the singlet states of donor and acceptor provides values for the transfer rate, $k_{A*B}$, of $1{\cdot}7 \times 10^{11}$ l mol$^{-1}$ s$^{-1}$ in benzene and $1{\cdot}4 \times 10^{11}$ l mol$^{-1}$ s$^{-1}$ in liquid paraffin. Similar rates were found with other donor–acceptor pairs in solution and with the 1-chloroanthracene–perylene pair in glassy chloroalkane solvents at 90 K. The latter data together with those for liquid paraffin provide compelling support for a long-range process.

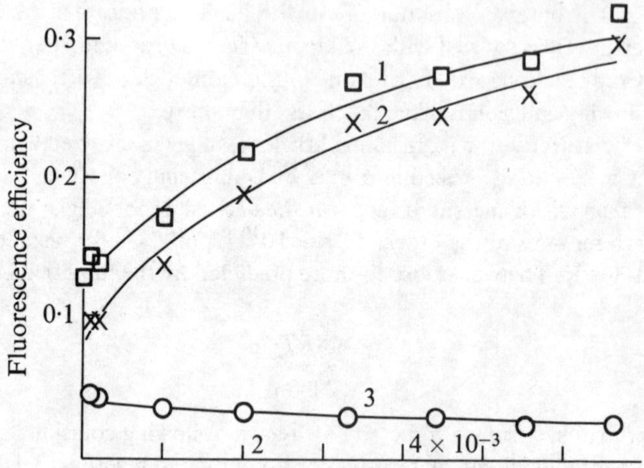


**Fig. 3.1** Fluorescence of mixed solutions in benzene of 1-chloroanthracene and perylene in 5:1 molar ratio, curve 1– mixed solution; curve 2—perylene alone; curve 3—1-chloroanthracene alone[4]. (Reproduced with permission of the authors and publisher.)

A theoretical basis for discussion of transfer of this type has been provided by Förster[6] who obtains the following expression for $k_{A*B}$

$$k_{A*B} = \frac{9000\,(\ln 10)\,\kappa^2\,\varphi_A}{128\pi^5\,n^4\,N\tau_A\,R^6} \int_0^\infty f_A(\bar{\nu})\,\epsilon_B(\bar{\nu})\,\frac{d\bar{\nu}}{\bar{\nu}_4} \qquad (3.4)$$

$\kappa$ is an orientation factor, e.g., $(2/3)^{1/2}$ for a random distribution, $\varphi_A$ and $\tau_A$ are the fluorescence quantum yield and mean lifetime of the donor respectively, $n$ is the refractive index of the solvent, $N$ is Avogadro's number, $R$ is the average separation of A and B, $f_A(\bar{\nu})$ is the fluorescence spectrum of the donor A, $\epsilon_B$ is the molar decadic extinction of the acceptor B at wave number $\bar{\nu}$. A related expression exists for the distance $R_0$ at which spontaneous decay of A* and transfer to B are equally probable,

$$R_0{}^6 = \frac{9000\,(\ln 10)\,\kappa^2\,\varphi_A}{128\,\pi^5\,n^4\,N\bar{\nu}^4} \int_0^\infty f_A(\bar{\nu})\,\epsilon_A(\bar{\nu})\,d\bar{\nu} \qquad (3.5)$$

Values of $R_0$ of 5–10 nm are predicted from this equation and $k_{A^*B}$ is found to be greatly in excess of the collision number $Z_{A^*B}$. The satisfactory agreement between the Förster model, which is based on a dipole-dipole interaction, and experimentally determined transfer distances is exemplified by Bennett's results[7] with, for example, pyrene (donor) and the dye Sevron yellow GL (acceptor) in solution in a rigid cellulose acetate film; here $R_0$ (theoretical) = 3·9 nm and $R_0$ (found) ~ 4·2 nm.

Other examples of energy transfer in fluid and non-crystalline rigid solution involve changes in multiplicity of donor (D) and acceptor (A). The various combinations can be summarized,

$$^3D^* + {}^1A \rightarrow {}^1D + {}^3A^* \tag{3.6}$$
$$^3D^* + {}^3A^* \rightarrow \text{singlet} + {}^1A^* \tag{3.7}$$
$$^3D^* + {}^3A \rightarrow {}^1D + {}^3A^* \tag{3.8}$$
$$^3D^* + {}^1A \rightarrow {}^1D + {}^1A^* \tag{3.9}$$
$$^3D^* + {}^3A \rightarrow {}^1D + {}^3A^* \tag{3.10}$$

Sensitized phosphorescence, corresponding to equation (3.6), was demonstrated by Terenin and Ermolaev[8] using rigid, glassy solutions at low temperatures of various aromatic carbonyl compounds as donors, and naphthalene or one of its derivatives as acceptor. Excitation of benzophenone to $S_1$ is followed by intersystem crossing to the first triplet level $T_1$ and either benzophenone phosphorescence or transfer to give triplet naphthalene followed by naphthalene phosphorescence (Fig. 3.2). In conformity with this scheme, successive addition of naphthalene reduced both the intensity and lifetime of the benzophenone phosphorescence and enhanced the intensity of naphthalene phosphorescence. Transfer distances of the order of 1·1–1·5 nm obtain with this mode of transfer, indicating that short-range exchange forces are responsible.

So far, all examples considered have involved solutions of both donor and acceptor in either a fluid solution or a glassy rigid medium, and the intervening solvent molecules are not regarded as participating in the mechanism of transfer. An alternative situation obtains when the donor is in the form of a (host) crystal containing deliberately added trace quantities of the acceptor, denoted the 'guest' molecule. Both singlet–singlet and triplet–triplet transfer have been demonstrated in such molecular crystals, mainly by luminescence studies[9]. Molecules capable of fluorescing in dilute solution also fluoresce as molecular crystals, to give in some cases a rather similar spectrum but in others, for example pyrene, a broader, structureless spectrum similar to that of the pyrene 'excimer' (p. 61). Again, naphthalene crystals show a structureless fluorescence attributed to emission from an excited state delocalized over a group of molecules and designated a singlet 'exciton'. This may be regarded as a band of energy states obtained by combining the wave functions of $S_1$ levels of a group of naphthalene molecules in a condition of high local order (Fig. 3.3).[10]

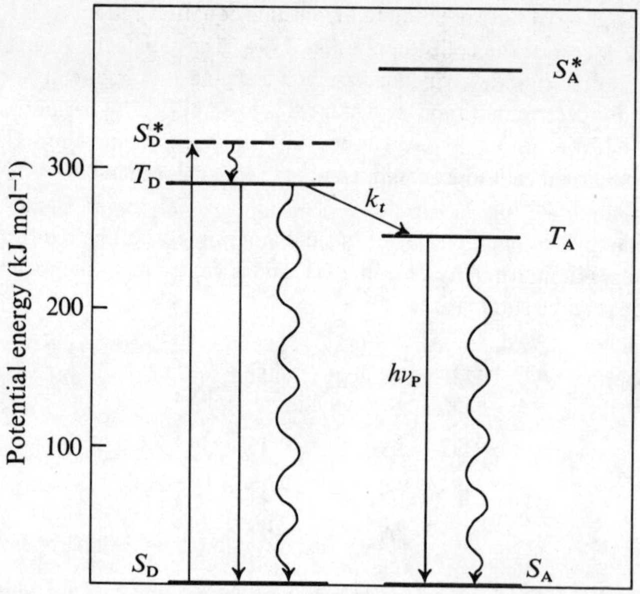


**Fig. 3.2** Electronic levels of the energy donor and acceptor in the phenomenon of sensitized phosphorescence (relative height for benzophenone* + naphthalene) (Reproduced with permission of the authors and publisher.[8])

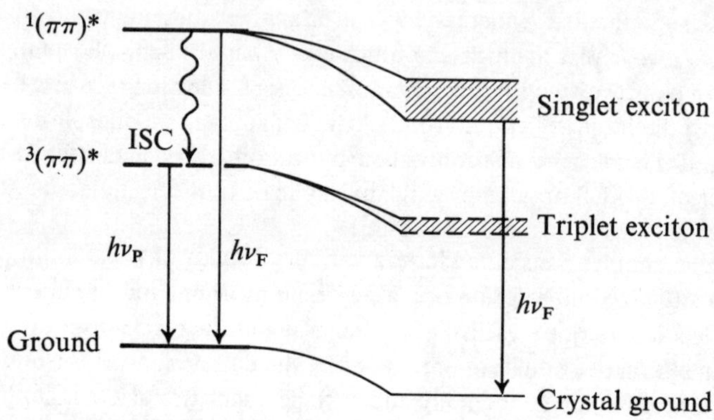


**Fig. 3.3** Effect of crystallization on molecular singlet and triplet states and on known processes. (Reproduced with permission of the author and publisher.[10])

The delocalization of singlet energy is apparent on adding phenanthrene as a guest at $10^{-4}$ mole concentration per mole of naphthalene when the emission consists of equal parts of naphthalene and phenanthrene fluorescence, indicating an efficiency of singlet transfer quite peculiar to the crystalline state. In a different kind of experiment a series of layers of pure anthracene of varying thickness were prepared on top of which were placed 'detector' layers of naphthacene (1 part) in anthracene (300 parts). Light was directed into the pure anthracene layer and naphthacene emission from the mixed crystal layer monitored. The diffusion length of the exciton was found to be 45 nm.

Evidence is accumulating that excitons participate in liquid aromatic systems; for example transfer rates from benzene (donor) to *p*-terphenyl on radiolysis or photolysis exceed those expected for diffusion control. In solution, of course, the degree of order must be small and is presumably confined to very localized regions.

The distinction between the cases of isolated molecules and crystalline aggregrates becomes particularly clear for triplet state excitation.[10] Whilst naphthalene phosphoresces in rigid solution at 77 K ($\varphi_P = 0{\cdot}03$), crystals of naphthalene and of other aromatic hydrocarbons and also mixed crystals of hydrocarbons phosphoresce only extremely weakly even at 4·2 K. The fate of the triplet states is either to engage in mutual annihilation (equation (2.2)) or to become deactivated by a more efficient crossover to $(S_0)^v$ or by localization of the exciton at a lattice defect followed by exchange of electronic energy with lattice states. Although delayed fluorescence can be detected from anthracene crystals excited by a laser source, this constitutes a minor path under normal conditions of photoexcitation; however, the laser experiments indicate the major path of triplet decay to be unimolecular.

The behaviour of crystals of aromatic ketones offers a complete contrast to that of polyacenes. These display strong phosphorescence in yields comparable with isolated molecules and of similar spectral character. Transfer to guest impurities, e.g., benzanthracene, takes place over larger distances than for *singlet* energy transfer in polyacene crystals, and a value for the diffusion length of the triplet exciton in anthracene of 5000 nm has been obtained. The dichotomy between polyacenes and ketones as regards crystal phosphorescence lies in the $(n, \pi)^*$ character of the $T_1$ state of the latter which implies an extremely short lifetime ($\sim 10^{-3}$s compared with 10 s for a polyacene crystal), and hence a correspondingly reduced probability of transitions into lattice sites as heat.

While discussion has been confined to energy transfer following photoirradiation, this process in its various forms is, of course, highly relevant to any understanding of radiation-chemical action, particularly as regards aromatic liquids and solutions, polymers and organic and inorganic crystals. Transfer in the former systems depends on processes already discussed, but energy transfer in inorganic crystals depends on a different kind of exciton, which is illustrated by zinc sulphide, and which although self-luminescent, is copper-activated for use in cathode-ray tubes. Impact of an electron beam promotes electrons to both the conduction band and to 'traps'. The electron may then move through the band until it returns to a $Cu^+$ ion to give the green emission. 'Trapped' electrons may be released by warming the solid or by photoactivation. An alternative model envisages occupancy by the electron of a hole in the valence band in the region of interstitial copper. Release then produces the blue emission of copper-activated ZnS.

The foregoing discussion has concentrated upon the photophysical consequences of energy transfer. Equally significant are the chemical reactions of the acceptors (which are said to be photosensitized) which can involve either $^1A^*$, $^3A$ or dissociative states formed by singlet–singlet transfer and reaction (3.6) respectively. Sensitized reactions of several different types have been identified, and some examples have been given of those of Hg($^1P_1$) and ($^3P_1$) in section 2.1. Excited molecules can also sensitize both homolytic reactions, for example, that of anthracene with carbon tetrachloride,[11]

$$^3(C_{14}H_{10}) + CCl_4 \rightarrow C_{14}H_{10}Cl^{\cdot} + CCl_3\cdot$$

and the isomerization of quadricyclane to norbornadiene using triplet aromatic ketones as sensitizing agent (section 4.2.2), and also *cis-trans* isomerization, dimerization, and addition reactions of olefins, again largely through the agency of triplet aromatic ketones which are formed with a quantum efficiency approaching unity on irradiation. The latter group of reactions have stimulated a great deal of discussion of points of mechanistic detail, including the geometry of the excited olefin, and the relative positioning of donor singlet and acceptor triplet levels ($T_1$ and $T_2$), and a general summary is presented in section 4.2.1.

Quantum yields for isomerization of the acceptor olefins are a measure of that of the triplet donor, $\varphi_T$; examples of $\varphi_T$ are given in Table 3.1.

TABLE 3.1: $\varphi_T$ for donors in sensitized isomerization experiments

| Donor | λ (nm) | $\varphi_T$ | Acceptor | Reference |
|---|---|---|---|---|
| Benzene (vapour) | 253·7 | 0·63 ± 0·01 | *cis*-but-2-ene and *cis*-ethylene-$d^2$ | 12 |
| Benzene-$d^6$ (vapour) | 253·7 | 0·55 ± 0·01 | As above | 12 |
| Fluorobenzene (vapour) | 248·0, 253·7, 265·0 | 0·86 ± 0·04 | *cis*-but-2-ene | 13 |
| Acetone (hexane solution) | 313·0 | 1·0 | *cis*- and *trans*-pent-2-ene | 14 |
| Naphthalene (benzene solution) | 313·0 | 0·40 ± 0·01 | *cis*-piperylene and *trans*-β-methylstyrene | 15 |
| Phenanthrene (benzene solution) | 313·0 | 0·76 ± 0·01 | As above | 15 |
| Fluorene (benzene) | 313·0 | 0·31 ± 0·01 | As above | 15 |
| Triphenylene (benzene) | 313·0 | 0·95 ± 0·05 | As above | 15 |

## 3.2 Simple reactions of stable* singlet and triplet states

A number of molecules undergo exceptionally simple reactions whilst in their excited singlet or triplet states, namely:

(*a*) Interaction with a second, unexcited molecule of identical or closely similar character to give a 'dimeric' excited state or 'excimer',

(*b*) Interaction with a different molecule to give a *transient* charge-transfer

* i.e., non-dissociative

complex or 'exciplex' which may dissociate to a radical-cation–radical-anion pair,

(*c*) Ionization of a proton,

(*d*) Hydrogen atom abstraction,

and some examples of each are given below. The extrusion of simple molecules such as nitrogen, hydrogen, and carbon monoxide is covered in section 4.1.

### 3.2.1 Excimer formation

Although $10^{-6}$ M solutions of pyrene in ethanol exhibit typical, vibrationally structured fluorescence spectra in the region 360–400 nm, the spectra of stronger solutions show additional emission until at $3 \times 10^{-3}$ M solute concentration a broad structureless band ($\lambda_{max}$ 480 nm) dominates the luminescence, originating from a dimer formed by interaction of a singlet excited molecule $S_1$ with an unexcited molecule $S_0$

$$S_1 + S_0 \rightleftharpoons (S_0\, S_1)^* \rightarrow 2S_0 + h\nu_{F(Ex.)}$$

It is important to appreciate that dimerization of two $S_0$ molecules prior to excitation does not occur and there is no parallel complication of the absorption spectrum. The bonding in an excimer probably involves contributions from both charge-transfer states and the delocalization of singlet energy between two $\pi$-systems, but their relative importance cannot be assessed at present. Excimer formation is more general than originally supposed and *absorption* by an excimer has recently been detected in liquid benzene ($\lambda_{max}$ 515 nm) by means of ns pulse radiolysis.

### 3.2.2 Transient charge-transfer complexes

We have referred on p. 41 to the photoexcitation of charge-transfer (CT) complexes. Certain donors, whilst unable to form a CT complex with a given acceptor under normal conditions, can do so on photoexcitation to give an excited CT complex, also known as an 'exciplex' or a 'heteroexcimer'

$$D(S_0) \xrightarrow{h\nu} D(S_1)$$

$$D(S_1) + A \rightleftharpoons (D.A)^*$$

Alternatively, in some cases excitation of the acceptor leads to formation of (DA)*, which may then undergo either typical photophysical processes, such as fluorescence, or dissociation,[16]

$$(D.A)^* \rightarrow D^{\dot{+}} + A^{\dot{-}}$$

A given CT complex may fluoresce in non-polar solvents and dissociate in polar solvents.

CT fluorescence is normally broad and structureless, appearing at longer wavelengths than the donor fluorescence. A typical pair engaging in this type of interaction are N,N,N′,N′-tetramethyl-*p*-phenylenediamine (TMPD, donor) and 1-methylnaphthalene (acceptor).[17] In alkane solvents at room temperature TMPD fluoresces at 390 nm but progressive introduction of 1-methylnaphthalene results in the gradual appearance of a new fluorescence at 480 nm at the expense of the 390 nm emission. No ground-state CT complex exists between this pair and the CT fluorescence is reduced when viscous solvents are used, indicating the need for collision between acceptor and singlet excited donor within the lifetime of the latter.

The dissociation step can be monitored in the case of N,N-dimethyl-naphthylamine (excited donor) and dimethyl isophthalate by means of flash photolysis in N,N-dimethylformamide solution which produces the spectrum of $D^{+\cdot}$, the donor radical-cation.[18]

### 3.2.3 Ionization

Acids (HA) and bases ($BH^+$) can undergo remarkable changes, of up to 8 $pK_a$ units, in their acidity on excitation to the $S_1$ state, which accounts for changes of the fluorescence spectra with pH of certain molecules *not* parallelled in the corresponding absorption spectra. $\Delta pK_a = pK_a - pK_a^*$ is related to the difference in the electronic frequencies in $cm^{-1}$ of the acid HA and its conjugate base $A^-$ by means of a thermochemical cycle, e.g., for a phenol, ArOH; this situation can be represented as in Fig. 3.4.

Clearly
$$\Delta E_{HA} + \Delta E_D^* = \Delta E_{A^-} + \Delta E_D$$

where $\Delta E_D$, $\Delta E_D^*$ refer to the energies of dissociation of ArOH and ArOH* respectively, and

$$\Delta E_D - \Delta E_D^* = (\Delta G_D + T\Delta S_D) - (\Delta G_D^* + T\Delta S_D^*)$$

Assuming
$$\Delta S_D = \Delta S_D^*$$

then
$$\begin{aligned}\Delta G_D - \Delta G_D^* = -\mathbf{R}T(\ln K_a - \ln K_a^*) &= \Delta E_D - \Delta E_D^* \\ &= \Delta E_{HA} - \Delta E_{A^-} \\ &= hc\bar{\nu}_{HA} - hc\bar{\nu}_{A^-}\end{aligned}$$

i.e.,
$$pK_a - pK_a^* = \frac{hc}{2{\cdot}303\mathbf{k}T}(\bar{\nu}_{HA} - \bar{\nu}_{A^-})$$

In general, determination of $pK_a^*$ from fluorescence 'titration' curves gives results in good agreement with those calculated from spectroscopic properties, indicating phenols and aromatic amine cations, $ArNH_3^+$, to become more acidic on excitation by four to seven $pK_a$ units. On the other hand, singlet excited benzoic acid is *more* protonated (as $C_6H_5CO_2H_2^+$) than the ground-state molecule by about 7 $pK_a$ units.

The origin of $pK_a$ changes on excitation lies in changes in charge distribution, for example, promotion of an electron from the oxygen atom of phenol to the ring in a $\pi^* \leftarrow n$ transition leads to a more positive oxygen atom in $^1(C_6H_5OH)^*$

and consequently greater acidity. The charge densities can be obtained by MO calculations,[19] e.g., with phenol

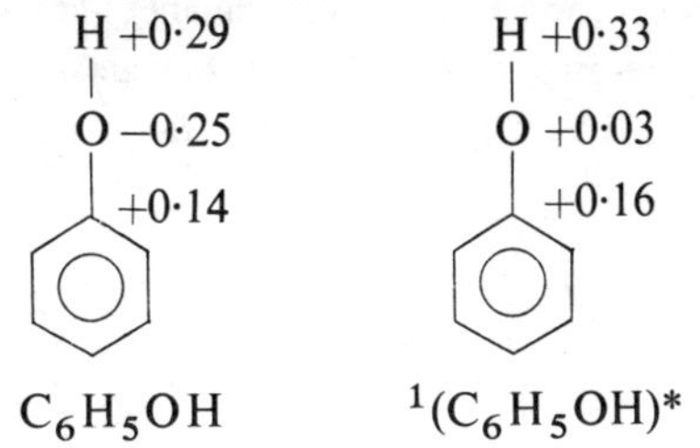


Analogous, but much smaller differences between the p$K_a$ values of a

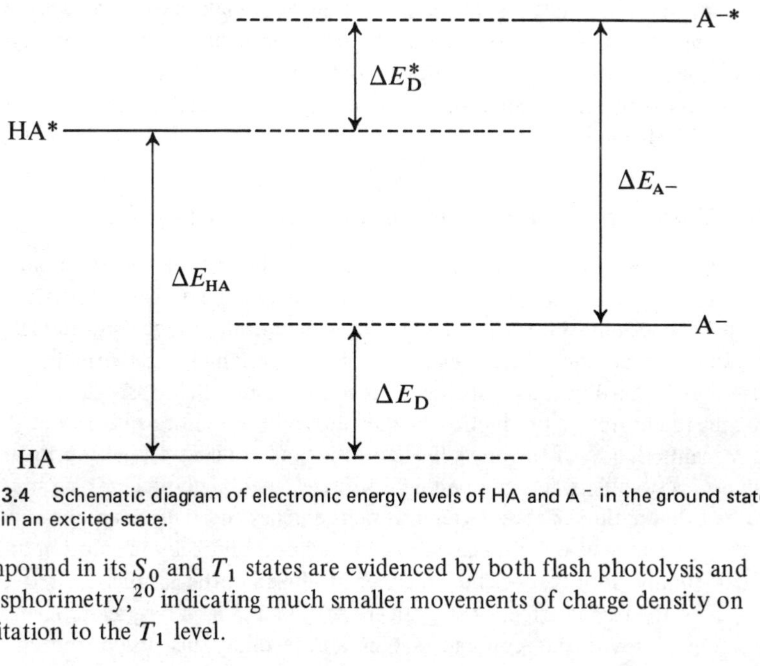


**Fig. 3.4** Schematic diagram of electronic energy levels of HA and $A^-$ in the ground state and in an excited state.

compound in its $S_0$ and $T_1$ states are evidenced by both flash photolysis and phosphorimetry,[20] indicating much smaller movements of charge density on excitation to the $T_1$ level.

### *3.2.4 Hydrogen atom abstraction*

The reduction of quinones and ketones to quinols and pinacols respectively on photolysis in hydrogen-containing solvents, including alkanes and alcohols, is sure indication of an abstraction from solvent. Designation of the abstracting species as the triplet state in the case of benzophenone follows from Porter and Wilkinson's[21] demonstration of the protecting effect of small added amounts of naphthalene, which has a lower $T_1$ level but a *higher* $S_1$ level than benzophenone (Fig. 3.2), upon the photolysis of benzophenone in propan-2-ol at wavelengths absorbed only by the ketone. The acceptance of energy from the reactive $^3(n, \pi^*)$ state of benzophenone by the unreactive $^3(\pi, \pi^*)$ state of naphthalene was also evident from the $\mu s$ flash photolysis spectra of the mixed solution, when the absorption of the ketyl radical $(C_6H_5)_2\dot{C}OH$ ($\lambda_{max}$ 545 nm) was progressively reduced on addition of naphthalene whilst that of the $T_1$ state of the latter ($\lambda_{max}$ 415 nm) increased in proportion. The processes taking place can be summarized

$$Ph_2CO \xrightarrow{h\nu} {}^1(Ph_2CO)^* \xrightarrow{\varphi \sim 1} {}^3(Ph_2CO)$$

$$^3(Ph_2CO) \xrightarrow{i\text{-PrOH}} Ph_2\dot{C}OH + (CH_3)_2\dot{C}OH \qquad {}^3(Ph_2CO) \xrightarrow{C_{10}H_8} {}^3(C_{10}H_8)$$

Frequency-doubled ruby laser flash photolysis examination ($\lambda$ = 347 nm) of the quenching of $^3(Ph_2CO)$ in benzene by naphthalene affords a direct measurement[22] of the quenching rate constant of $(4{\cdot}7 \pm 0{\cdot}4) \times 10^9$ l mol$^{-1}$ s$^{-1}$.

The failure of certain substituted benzophenones to abstract hydrogen is discussed in section 4.4.1.

### *3.2.5 Reactions of simple radicals produced photolytically*

Photolysis of a compound may produce simple radicals either in ground or electronically excited states, and, on occasion, with a great deal of kinetic energy. The ground-state radicals will behave chemically exactly as thermally produced radicals and only the electronically, vibrationally, or thermally excited variety will produce distinctive reactions. Nonetheless, special interest attaches to photolytic production of even ground-state radicals in view of the ready manipulation of the prevailing conditions of concentration by utilizing high intensity, pulsed, or chopped light sources, enabling optical, e.s.r., and kinetic examination of these often transient entities and their reactions.

The environment of a radical pair produced photolytically in solution differs fundamentally from that of the same pair produced in the gas phase, firstly, because of the caging effect of the solvent, which implies a far greater chance of recombination with the geminate partner, and secondly, because of the enhanced possibilities of deactivation of electronic, vibrational and thermal energy of the radicals.

The cage effect is illustrated by measurement[23] of $\varphi(I_2)$ in the 435·8 nm photolysis of molecular iodine in the presence of allyl iodide and oxygen (given in Table 3.2). Dissociation of iodine into atoms $(^2P_{1/2})$ and $(^2P_{3/2})$ is followed by abstraction of an iodine atom from allyl iodide to give a molecule of iodide;

$$I_2 \xrightarrow{h\nu} I^{\cdot}(^2P_{3/2}) + I^{\cdot}(^2P_{1/2})$$

$$I^{\cdot} + CH_2{=}CH{-}CH_2I \rightarrow I_2 + \overline{CH_2{-}\dot{C}H{-}CH_2}$$

$$\underline{CH_2{-}\dot{C}H{-}CH_2} + O_2 \rightarrow \text{peroxy radical} \rightarrow \text{stable products}$$

The effect of increase of solvent density in reducing the yield of escaped iodine atoms is striking; the temperature dependence is less obviously explained but suggests the existence of an activation energy of some magnitude for escape from the solvent cage. Cage recombination is an important factor in reducing the efficiency of radical thermal and photoinitiators in vinyl polymerization below the theoretical figure.

TABLE 3.2: Quantum yields in the photodissociation of iodine[23]

| Solvent | Temperature (K) | $\varphi (I_2)$ |
|---|---|---|
| Hexane | 288 | 0·50 ± 0·04 |
| | 298 | 0·66 ± 0·04 |
| $CCl_4$ | 290·5 | 0·11 ± 0·01 |
| | 298 | 0·14 ± 0·01 |
| | 311 | 0·21 ± 0·02 |
| Hexachlorobuta-1,3-diene | 288 | 0·042 ± 0·006 |
| | 298 | 0·075 ± 0·009 |
| | 308 | 0·15 ± 0·02 |

The contrast between gas and liquid phases is further illustrated by 253·7 nm photolysis of acetone. The much lower quantum yields for decomposition in the neat liquid and in solution are explained both in terms of a solvent cage effect and of deactivation of excited states by the solvent (Table 3.3). Acetone photochemistry is discussed in detail in section 4.1.1.

TABLE 3.3: Quantum yields in acetone photolysis at 253·7 nm

| Conditions | $\varphi CH_4$ | $\varphi CO$ | $\varphi C_2H_6$ | Reference |
|---|---|---|---|---|
| Vapour | $10^{-3}$ | 0·25 | 0·35 | 24 |
| Liquid | $\sim 2 \times 10^{-3}$ | $\sim 10^{-4}$ | $\sim 10^{-4}$ | 25 |
| Aqueous solution (0·108 M) | $3{\cdot}6 \times 10^{-2}$ | $2{\cdot}2 \times 10^{-3}$ | $2{\cdot}9 \times 10^{-2}$ | 26 |
| Perfluorodimethyl–cyclobutane solution (0·096 M) | $3{\cdot}5 \times 10^{-4}$ | $2{\cdot}6 \times 10^{-4}$ | $2{\cdot}5 \times 10^{-4}$ | 27 |

The variety of reactions instigated by simple radicals produced photochemically is apparent just from a consideration of atoms. A classical reaction is that of hydrogen and chlorine which shows all the characteristics of a chain process including a quantum yield of up to $10^5$, a dependence on the half-power of the intensity of illumination, sensitivity to traces of water, oxygen and the nature of the surface of the vessel and the appearance of induction periods, i.e., quiescence before onset of reaction, on adding ammonia. The reaction scheme in oxygen-free systems is[28]

$$Cl_2 + h\nu(406\ \text{nm}) \xrightarrow{k_1} 2Cl^{\bullet}(^2P_{1/2});\ \text{intensity} = I_a \tag{3.11}$$

$$Cl^{\bullet} + H_2 \xrightarrow{k_2} HCl + H^{\bullet} \tag{3.12}$$

$$H^{\bullet} + Cl_2 \xrightarrow{k_3} HCl + Cl^{\bullet} \tag{3.13}$$

$$H^{\bullet} + HCl \xrightarrow{k_{-2}} H_2 + Cl^{\bullet} \tag{3.14}$$

$$2Cl^{\bullet} \xrightarrow{k_4} Cl_2 \tag{3.15}$$

and a stationary-state treatment leads to an overall rate given by

$$\frac{d[HCl]}{dt} = \frac{2k_2 k_3 (k_1/k_4)^{1/2}[H_2][Cl_2]}{(k_3[Cl_2] + k_{-2}[HCl])} I_a^{1/2} \tag{3.16}$$

where $k_{-2}/k_3 = 1{\cdot}7$ and $2k_2(k_1/k_4)^{1/2} = 1{\cdot}8 \times 10^{-3}$. Equation 3.15 is highly exothermic and a wall reaction or a transient $Cl_3^{\bullet}$ species has been suggested as the means of dissipating excess energy in view of the pressure-independence of $\varphi_{HCl}$.

The resolution of expressions such as equation 3.16 into individual estimates of the rate constants for the component parts depends on the application of flash photolysis or rotating sector techniques. The latter is applied exclusively to systems with a kinetic dependence on a fractional power of $I_a$ and is based on the chopping of the incident light beam by means of a rapidly rotating circular metal disc from which sectors have been cut. The rotation rate can be up to 35 revolutions per second and the flashing rate is of higher frequency still, depending on the number of sectors cut out from the disc. If, for example, these number four, have angles of 22·5°, and are evenly spaced then the light and dark periods 'seen' by the vessel will be in a ratio of one to three and the flashing rate will be four times the rotation frequency of the disc. For a generalized chain reaction,

$$X_2 + h\nu \xrightarrow{k_1} 2X^{\bullet};\ \text{intensity} = I_a,\ \text{quantum yield} = \varphi$$

$$X^{\bullet} + A \xrightarrow{k_2} B + Y^{\bullet}$$

$$Y^{\bullet} + X_2 \xrightarrow{k_3} XY + X^{\bullet}$$

$$2X^{\bullet} \xrightarrow{k_4} X_2$$

a stationary-state treatment gives

$$[X^{\bullet}] = \left(\frac{\varphi I_a}{2k_4}\right)^{1/2} \tag{3.17}$$

At slow rotation speeds, the effect of the disc is to cause the reaction to proceed in a sequence of on-off periods (of relative duration 1:3 for the disc described) under the same intensity of illumination as in the complete absence of the disc, $I_a$. At speeds sufficiently high to maintain the reaction in a steady-state condition, that is, such that the disappearance of chain carriers $X^{\bullet}$ during the dark periods is negligible, then the steady-state concentrations of $X^{\bullet}$ is given by

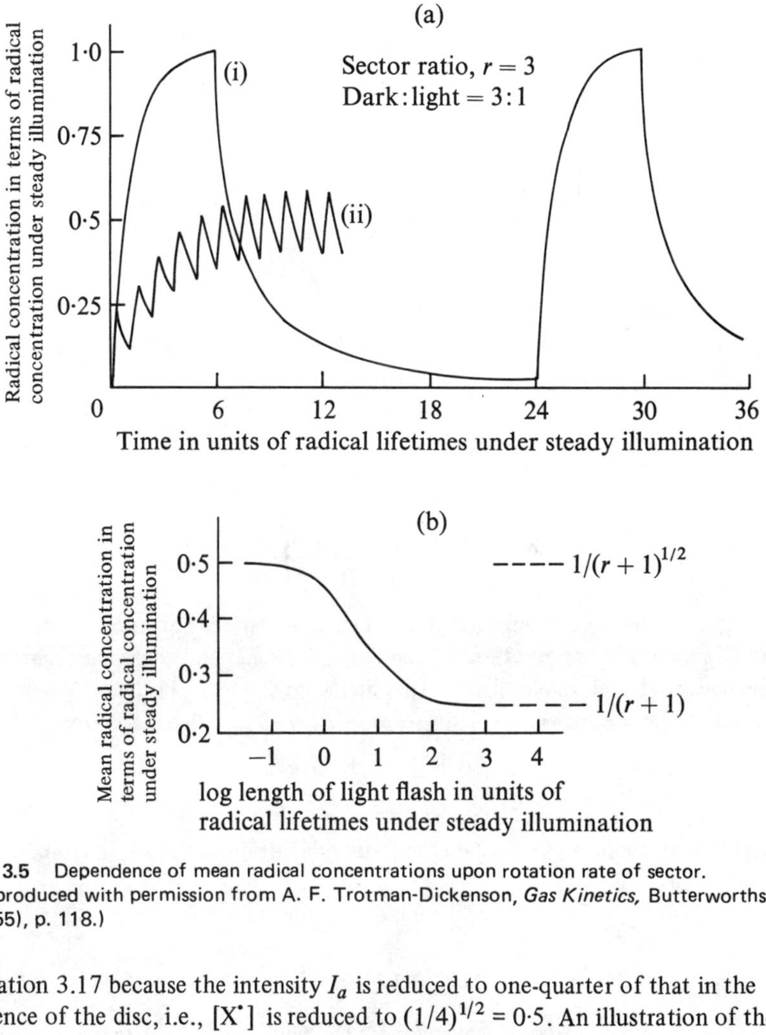


Fig. 3.5 Dependence of mean radical concentrations upon rotation rate of sector. Reproduced with permission from A. F. Trotman-Dickenson, *Gas Kinetics*, Butterworths 1955), p. 118.)

equation 3.17 because the intensity $I_a$ is reduced to one-quarter of that in the absence of the disc, i.e., $[X^{\bullet}]$ is reduced to $(1/4)^{1/2} = 0{\cdot}5$. An illustration of the 'slow' and 'fast' cases is given in Fig. 3.5 and as the speed of rotation is increased it is clear that the overall reaction rate will be increased. The transition between the two extreme cases will occur in a region when $X^{\bullet}$ is disappearing appreciably, but not entirely, during the dark period, and a rather complex kinetic analysis affords values for $k_4$ in equation 3.17.

The method has found wide application in both simple gas phase and solution photopolymerization studies. The Bodenstein mechanism for the photosynthesis of phosgene is

$$Cl_2 + h\nu \xrightarrow{k_1} 2Cl^\bullet \qquad \varphi I_a$$

$$Cl^\bullet + CO \xrightarrow{k_2} COCl^\bullet$$

$$COCl^\bullet \xrightarrow{k_{-2}} Cl^\bullet + CO$$

$$COCl^\bullet + Cl_2 \xrightarrow{k_3} COCl_2 + Cl^\bullet$$

$$COCl^\bullet + Cl^\bullet \xrightarrow{k_4} CO + Cl_2$$

and the observed rate law is

$$\frac{d[COCl_2]}{dt} = k_3\left(\frac{k_1 k_{-2}}{k_2 k_4}\right)^{1/2} (\varphi I_a)^{1/2}[CO]^{1/2}[Cl_2]$$

which implies $k_{-2}\ [COCl^\bullet] \gg k_3\, [COCl^\bullet]\, [Cl_2]$. A rotating sector study[29] produced values for individual rate constants of

$$k_3 = 2{\cdot}5 \times 10^9 \exp\left(\frac{-12{\cdot}38\ \text{kJ}}{\mathbf{R}T}\right) \text{l mol}^{-1}\ \text{s}^{-1}$$

$$k_4 = 4{\cdot}0 \times 10^{11} \exp\left(\frac{-3{\cdot}47\ \text{kJ}}{\mathbf{R}T}\right) \text{l mol}^{-1}\ \text{s}^{-1}$$

Analogous studies have been carried out on a great variety of chain systems including the photochlorinations of alkanes and alkenes and their halogenated derivatives, and Bodenstein mechanisms similar to steps (3.11) to (3.15) are believed to operate for alkanes with steps (3.12) and (3.13) replaced by

$$Cl^\bullet + RH \rightleftharpoons R^\bullet + HCl$$

$$R^\bullet + Cl_2 \rightleftharpoons RCl + Cl^\bullet$$

Reaction with alkenes (A) involves preliminary addition of $Cl^\bullet$ to substrate in the sequence,

$$Cl^\bullet + A \rightleftharpoons ACl^\bullet$$

$$ACl^\bullet + Cl_2 \rightleftharpoons ACl_2 + Cl^\bullet$$

$$2\ ACl^\bullet \rightarrow \text{products } (A_2Cl_2 \text{ and/or } A + ACl_2)$$

$$ACl^\bullet + Cl^\bullet \rightarrow ACl_2$$

Conventional kinetic measurements of the radical polymerization of vinyl monomers produces a ratio for rate constants of propagation ($k_p$) and terminatio[n] ($k_t$) of $k_p/(k_t)^{1/2}$ but application of the rotating sector technique to photoinitiated polymerization provides the additional ratio $k_p/k_t$ which enables estimation of $k_p$ and $k_t$ separately, some typical values of which are given in Table 3.4.

TABLE 3.4: Rate constants for propagation and termination steps in vinyl photopolymerization at 330 K by rotating sector method

| Monomer | $k_p$ (l mol$^{-1}$ s$^{-1}$) | $10^{-7}$ $k_t$ (l mol$^{-1}$ s$^{-1}$) |
|---|---|---|
| Vinyl acetate | 2300 | 2·9 |
| Styrene | 145 | 0·13 |
| Methyl methacrylate | 705 | 1·8 |
| Vinyl chloride | 12 300 | 2300 |

O($^3P$) and ($^1D$) whilst not strictly free radicals, can be produced by photolysis of $NO_2^\bullet$ at $\lambda < 370$ and 228·8 nm respectively (Table 2.2). O($^3P$) can also be produced by the action of a microwave discharge upon $O_2$ and by Hg($^3P_1$) attack on $N_2O$ (Table 2.1), but thermal production is difficult. Attack of O($^3P$) upon olefins[30] to yield largely epoxides and carbonyl compounds is non-stereospecific, indicating retention of total spin to give initial adducts of the type

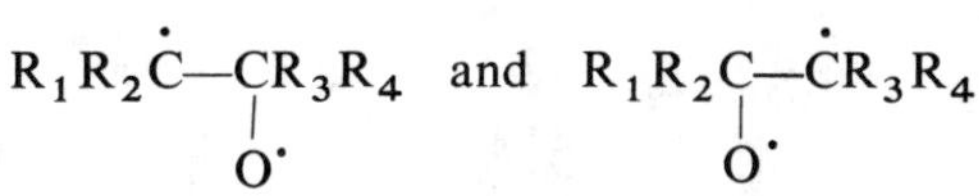

but the oxygen attaches preferentially, if not exclusively, to the less substituted carbon atom. O($^1D$) gives a *singlet* adduct in the form of an excited epoxide produced stereospecifically, but a competing path is collisional deactivation to O($^3P$).

Compilations of relative and absolute rate constants for O($^3P$) attack on olefins have appeared. For ethylene, which is typical, the absolute rate constant is given by

$$k_{298\,\mathrm{K}} = 8{\cdot}4 \times 10^9 \exp\left(\frac{-6{\cdot}7\ \mathrm{kJ}}{\mathbf{R}T}\right) \mathrm{l\,mol^{-1}\,s^{-1}}$$

Photodissociation of molecules such as HBr, HI, alkyl iodides, etc., results in the appearance of the excess energy as kinetic energy, for example, 253·7 nm photolysis of HI yields *ca.* 172 kJ mol$^{-1}$. Since momentum must be conserved between the fragments,

$$m_H v_H = m_I v_I$$

the ratio of the kinetic energies ($KE$) is

$$\frac{KE_H}{KE_I} \frac{m_H v_H^2}{2} \cdot \frac{2}{m_I v_I^2} = \frac{v_H}{v_I} = \frac{m_I}{m_H} = 127$$

i.e., the great majority of the excess energy is taken up by the light fragment. The presence of $(D^\bullet)^*$ in the 253·7 nm photolysis of DI in the presence of excess ethane is registered by the appearance of products due to reactions (3.18) and (3.19)

$$(D^{\bullet})^* + C_2H_6 \rightarrow HD + C_2H_5^{\bullet} \qquad (3.18)$$

$$C_2H_5^{\bullet} + DI \rightarrow C_2H_5D + I^{\bullet} \qquad (3.19)$$

but attack of $(D^{\bullet})^*$ upon DI is unlikely as the latter is the minor component and $(D^{\bullet})^*$ will have reacted or moderated before encountering DI.[31] Thermalized

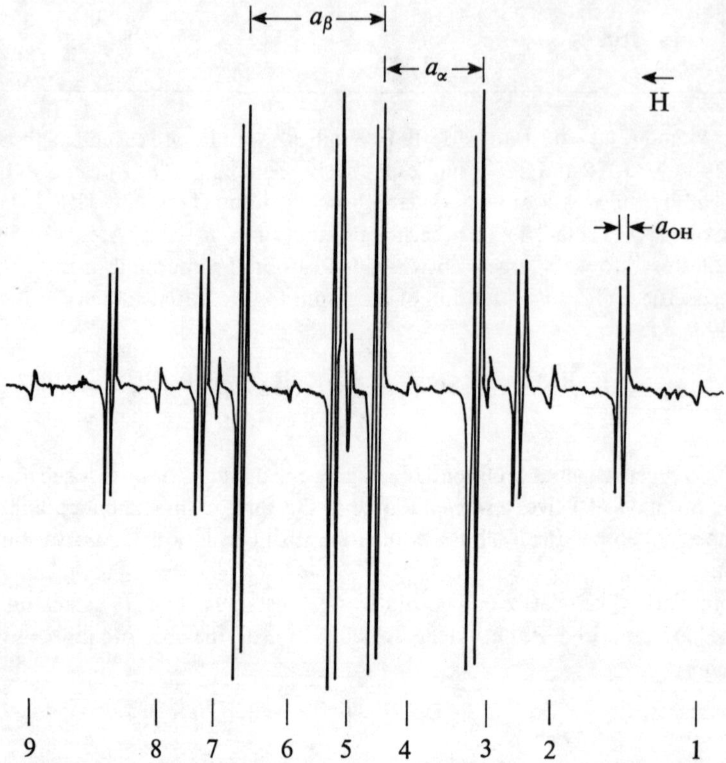


**Fig. 3.6** The spectrum formed by photolysis of ethyl alcohol containing $H_2O_2$ at 200 K. The numbered lines arise from $\cdot CH_2CH_2OH$ while the strong lines are from $CH_3\dot{C}HOH$. (Reproduced with permission of the authors and publisher.[32])

deuterium atoms $D^{\bullet}$ attack DI far more readily than $C_2H_6$, and the ratio $[HD]/[D_2]$ is accordingly a measure of $[(D^{\bullet})^*]/[D^{\bullet}]$. Addition of an inert gas reduces $[HD]/[D_2]$, indicating further moderation of $(D^{\bullet})^*$ by kinetic energy exchange.

'Hot' alkyl radicals are formed on addition of $H^{\bullet}$ to olefins,[30]

$$H^{\bullet} + C_2H_4 \rightarrow (C_2H_5^{\bullet})^*$$

although here the excess energy is vibrational in character. The reactions of such radicals are distinctive, for example $(C_2H_4D^{\bullet})^v$ can fragment to $H^{\bullet}$ and

$_2H_3D$, giving a misleading appearance of a bimolecular disproportionation of $_2H_5^{\bullet})^v$ as opposed to the normal recombination of thermalized $C_2H_5^{\bullet}$.

Ample evidence has been given of the power of photochemical methods in ɔtaining kinetic data for simple, fast reactions of radical species. It is also ıportant to recognize the wealth of spectroscopic information that has been ade available. In addition to vapour phase and solution studies of the u.v. ectra of simple radicals following flash photolysis and the i.r. and u.v. and s.r. work on radicals trapped in low-temperature matrices following adiation, there has recently been considerable development of interest[32,33] solution e.s.r. studies of simple radicals formed by attack on hydrogen-donor bstrates by photolytically generated oxidizing radicals, such as $OH^{\bullet}$ and $H_3)_3CO^{\bullet}$, and an example of a highly resolved spectrum so obtained is given in g. 3.6.

## eferences


. H. Ishikawa and W. A. Noyes, Jr., *J. Amer. Chem. Soc.*, **84**, 1502 (1962).
. J. T. Dubois and R. L. Van Hemert, *J. Chem. Phys.*, **40**, 923 (1964).
. J. Perrin, *Compt. Rend.*, **184**, 1097 (1924).
. E. J. Bowen and B. Brocklehurst, *Trans. Faraday Soc.*, **49**, 1131 (1953).
. E. J. Bowen and R. Livingston, *J. Amer. Chem. Soc.*, **76**, 6300 (1954).
. Th. Förster, *Disc. Faraday Soc.*, **27**, 7 (1959).
. R. G. Bennett, *J. Chem. Phys.*, **41**, 3037 (1964).
. A. N. Terenin and V. A. Ermolaev, *Trans. Faraday Soc.*, **52**, 1042 (1956).
. E. J. Bowen and B. Brocklehurst, *Trans. Faraday Soc.*, **51**, 774 (1955).
. R. M. Hochstrasser, *Photochem. Photobiol.*, **3**, 299 (1964).
. S. Kusuhara and R. Hardwick, *J. Chem. Phys.*, **41**, 3943 (1964).
. R. B. Cundall and A. S. Davies, *Trans. Faraday. Soc.*, **62**, 1151 (1966).
. D. Phillips, *J. Phys. Chem.*, **71**, 1839 (1967).
. R. F. Borkman and D. F. Kearns, *J. Amer. Chem. Soc.*, **88**, 3467 (1966).
. A. A. Lamola and G. S. Hammond, *J. Chem. Phys.*, **43**, 2129 (1965).
. H. Leonhard and A. Weller, *Z. Phys. Chem. (Frankfurt)*, **29**, 277 (1961).
. N. Yamamoto, Y. Nakato, and H. Tsubomura, *Bull. Chem. Soc., Japan*, **40**, 451 (1967).
. M. Koizumi and H. Yamashita, *Z. Phys. Chem. (Frankfurt)*, **57**, 103 (1968).
. C. Sándorfy, *Canad. J. Chem.*, **31**, 439 (1953).
. G. Jackson and G. Porter, *Proc. Roy. Soc.*, **A260**, 13 (1961).
. G. Porter and F. Wilkinson, *Trans. Faraday Soc.*, **57**, 1686 (1961).
. G. Porter and M. R. Topp, *Proc. Roy. Soc.*, **A315**, 163 (1970).
. F. W. Lampe and R. M. Noyes, *J. Amer. Chem. Soc.*, **76**, 2140 (1954).
. D. S. Herr and W. A. Noyes, Jr., *J. Amer. Chem. Soc.*, **62**, 2052 (1940).
. R. Pieck and E. W. R. Steacie, *Canad. J. Chem.*, **33**, 1304 (1955).
. D. H. Volman and L. W. Swanson, *J. Amer. Chem. Soc.*, **82**, 4141 (1960).
. R. Doepker and G. J. Maines, *J. Amer. Chem. Soc.*, **83**, 294 (1961).
. M. Ritchie and R. G. W. Norrish, *Proc. Roy. Soc.*, **A140**, 99; 112 (1933).
. W. G. Burns and F. S. Dainton, *Trans. Faraday. Soc.*, **48**, 39 (1952).
. R. J. Cvetanovic, *Adv. Photochem.*, **1**, 115 (1963).
. R. J. Carter, W. H. Hamill, and R. R. Williams, Jr., *J. Amer. Chem. Soc.*, **77**, 6457 (1955).
. R. Livingston and H. Zeldes, *J. Amer. Chem. Soc.*, **88**, 4333 (1966).
. P. J. Krusic and J. K. Kochi, *J. Amer. Chem. Soc.*, **91**, 6161 (1969).

# 4. Reactions of species produced photochemically—II

In this chapter organic reactions involving elimination, rearrangement, addition, abstraction, and substitution processes are discussed.

## 4.1 Elimination reactions

A common radiation induced sequence is fragmentation followed by expulsion of some species from one or the other of these fragments. Processes involving concerted cleavage of two $\sigma$ bonds terminating at a single atom are known as cheletropic reactions.[1]

### *4.1.1 Elimination of carbon monoxide*

A general description of the ground and excited states of carbonyl groups can b conveniently given in terms of the particular case of the carbonyl group in formaldehyde. The MO's may be classified as symmetric or antisymmetric with respect to the molecular plane (Fig. 4.1), the $2p_x$ orbital of carbon and the $2p_x$ orbital of oxygen being antisymmetric, and the $1s$, $2s$, $2p_y$, and $2p_z$ orbitals of both carbon and oxygen being symmetric with respect to this plane. Considerin the carbon orbitals first, it is apparent that the $1s$ orbital is full and, therefore, non-bonding while the $2s$, $2p_y$, and $2p_z$ orbitals can form three $sp^2$ hybrids. Tw of these enter into bonding with hydrogen, leaving the third to participate in th carbon–oxygen $\sigma$-bond.

By contrast, the oxygen $2s$ orbital features much less strongly in the $\sigma$-framewc because its energy is considerably smaller than that of the corresponding carbor orbital. For example, the excitation of carbon, $(2s2p^3)^3P \leftarrow (2s^2 2p^2)^3P$, requir 900 kJ mol$^{-1}$ (9·34 eV) but the corresponding excitation on oxygen, $(2s2p^5)^3P \leftarrow (2s^2 2p^4)^3P$, requires 1510 kJ mol$^{-1}$ (15·66 eV). Although the oxygen atom appears to possess a pair of $sp$ hybrid orbitals, these orbitals have unequal $s$ character, the one forming the $\sigma$-bond to carbon having less $s$ charact than the non-bonding pair. The remaining $2p_y$ orbital does not enter into bond

Radiation of wavelength $\simeq$ 280 nm usually promotes an electron from the $2p_y$ oxygen orbitals to a $\pi^*$ orbital delocalized over both carbon and oxygen. This results in the oxygen becoming electron deficient and hence, electrophilic and, conversely, the carbon becoming nucleophilic.

Reactions involving the elimination of carbon monoxide are known for aldehydes and ketones and have been observed in both the gas and condensed phases. Probably acetone[2] has been subjected to more gas phase photolytic studies than any other compound, the main products being carbon monoxide,

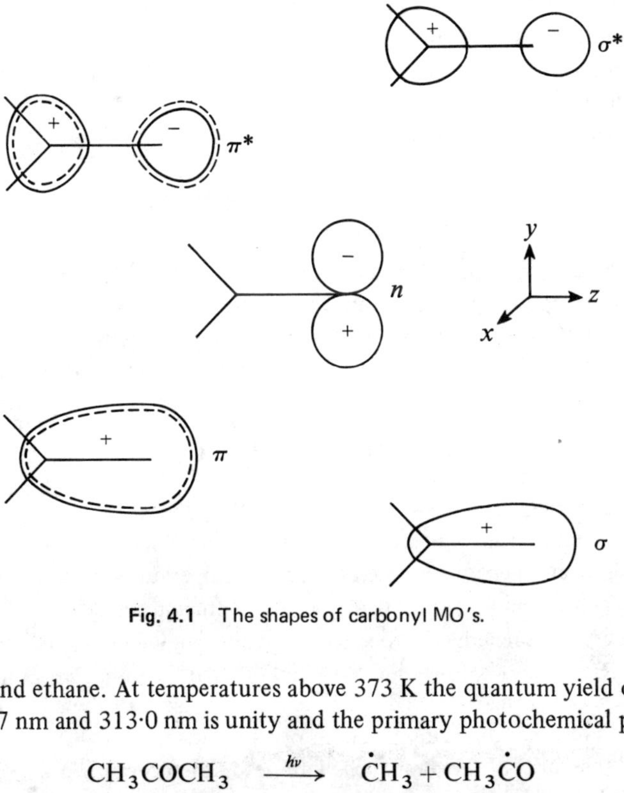


**Fig. 4.1** The shapes of carbonyl MO's.

methane and ethane. At temperatures above 373 K the quantum yield of CO at both 253·7 nm and 313·0 nm is unity and the primary photochemical process is

$$CH_3COCH_3 \xrightarrow{h\nu} \dot{C}H_3 + CH_3\dot{C}O$$

Elimination of CO from the acetyl radicals is a fast secondary process,

$$CH_3\dot{C}O \rightarrow \dot{C}H_3 + CO \qquad (4.1)$$

other secondary processes being atom abstraction,

$$\dot{C}H_3 + CH_3COCH_3 \rightarrow CH_4 + \dot{C}H_2COCH_3$$

and radical combination,

$$2\dot{C}H_3 \rightarrow C_2H_6$$

The quantum yields of methane and ethane vary in accordance with the mechanism as a function of reaction conditions.

Under some conditions ketene may be produced,

$$\dot{C}H_3 + CH_3\dot{C}O \rightarrow CH_4 + CH_2CO$$

along with biacetyl, methyl ethyl ketone, biacetonyl, and acetaldehyde.

The effects of environment on the photoprocesses of polyatomic molecules often produce profound changes in their photochemistry, due essentially to a combination of three factors.

(a) Vibrational deactivation by collision, a process largely absent in the vapour phase, is readily available in condensed phases.
(b) Bimolecular reaction with the solvent can lead to higher molecular weight products.
(c) A reaction may be quenched by operation of a cage effect.

$$R_2CO \underset{\text{fast}}{\overset{h\nu\ \text{solution}}{\rightleftarrows}} \underset{\text{solvent cage}}{[R^{\bullet} + R\dot{C}O]} \rightarrow 2R^{\bullet} + CO$$

For example, $\varphi_{CO}$ in pure liquid acetone at 298 K and 313·0 nm is 0·001, whereas, in the vapour under the same conditions it is 0·1. At 373 K, the quantum yield rises to unity and equation 4.1 becomes the most important secondary process. Consequently, at elevated temperatures the overall photochemical process is,

$$CH_3COCH_3 \rightarrow 2\dot{C}H_3 + CO$$

and the reaction then serves as a useful source of methyl radicals.

The excited state from which decomposition takes place can be either the singlet or triplet and depends on such factors as temperature, pressure, and wavelength, being predominantly triplet at 313·0 nm and singlet at 253·7 nm.

The primary photochemical process for acetone (shown above) in which the bond $\alpha$ to the carbonyl is cleaved, is a general mode of decomposition for ketones and is called the Norrish Type I Process. The Norrish Type II Process, or $\gamma$-hydrogen abstraction, is discussed on p. 136.

The photochemistry of several cyclic ketones in the vapour phase has been investigated. For example, photolysis of cyclopentanone[3] at 313·0 nm and at temperatures up to 398 K leads to four major products, probably via the $S_1$ state.

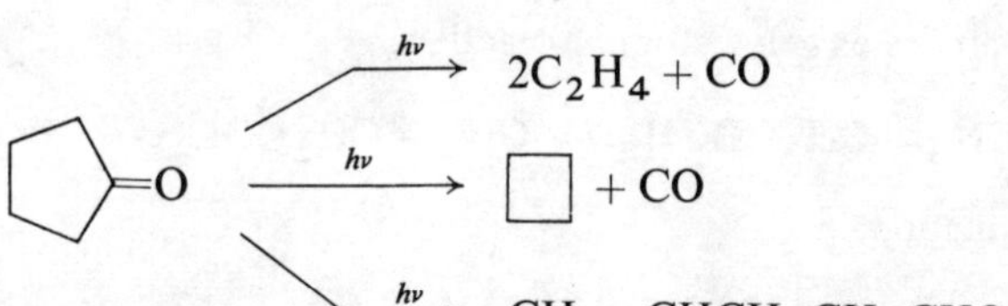

By analogy with simple aliphatic ketones, the primary photochemical process is assumed to be,

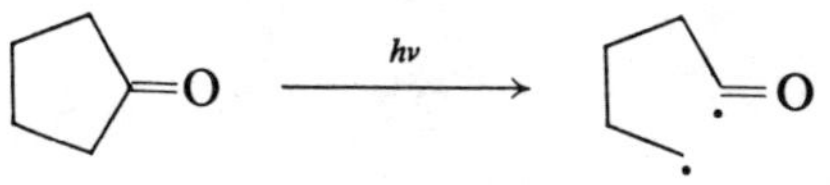


leading subsequently to,

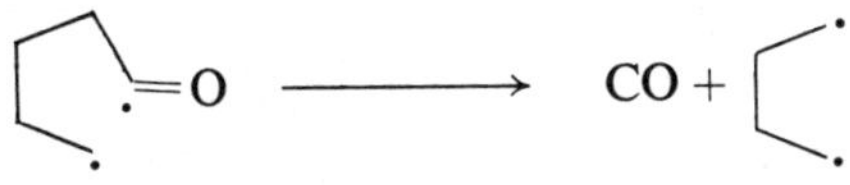


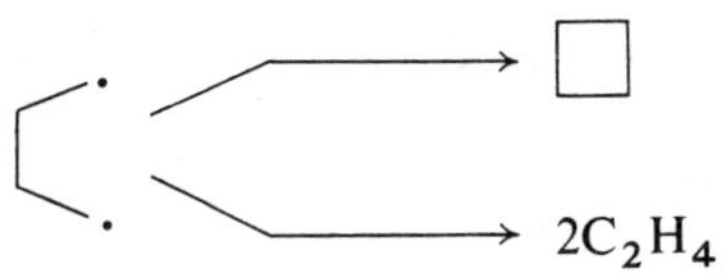


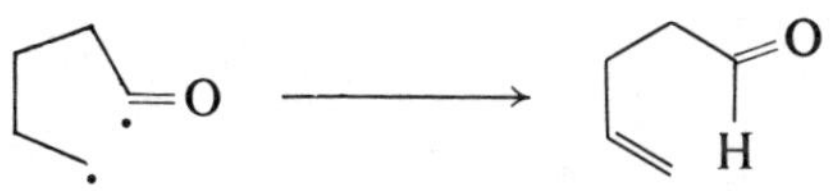


Elimination of carbon monoxide is well known in the condensed phase and is promoted by such factors as ring strain and radical stabilization. In some circumstances, it leads to structures only difficultly available by conventional chemical means, e.g.,

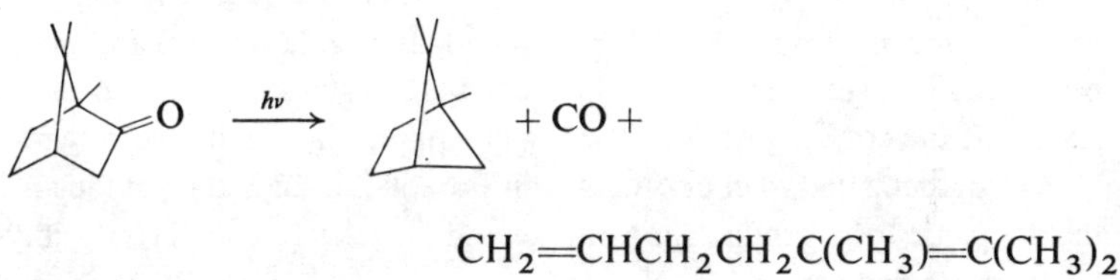


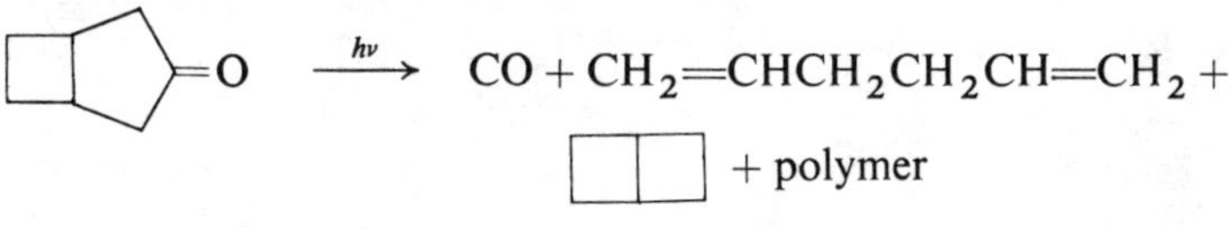


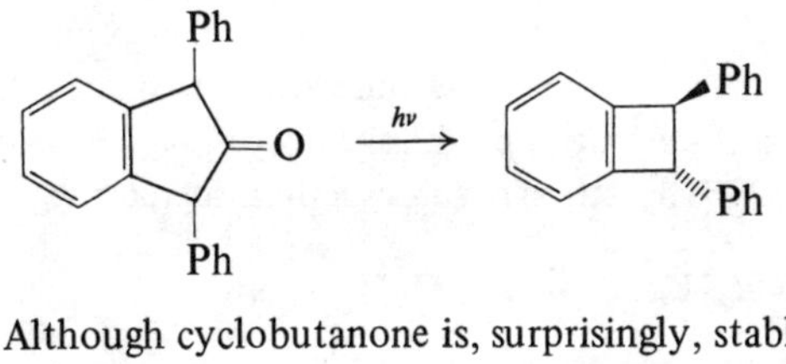


Although cyclobutanone is, surprisingly, stable to irradiation, tetramethylcyclobutadione in benzene solution is photolysed mainly to carbon monoxide and tetramethylethylene. The reaction proceeds through a diradical intermediate, which can be trapped as its furan adduct, to tetramethylcyclopropanone, which on loss of a further molecule of carbon monoxide gives tetramethylethylene.

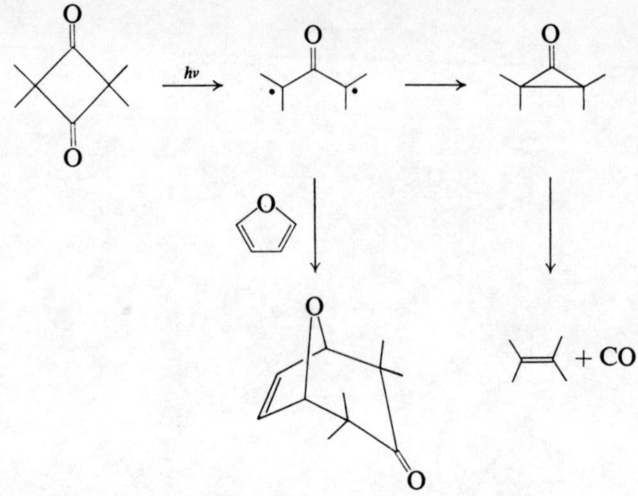


### 4.1.2 Elimination of nitrogen

Irradiation of aliphatic compounds containing the –N=N– link normally results in nitrogen elimination since this particularly facile process imparts a strong thermodynamic driving force for decomposition. Aromatic azo compounds undergo *cis-trans*-isomerisation (p. 86) and also oxidative cyclization to give benzcinnolines (cf. section 4.2.5). The lowest singlet is thought to be a $(n, \pi^*)$ state, but the configuration of the lowest triplet is generally unknown.

Azomethane has been photolysed in the solid, liquid, and gaseous phases. In the gas phase[4] the products are ethane and nitrogen ($\varphi_{N_2} \simeq 1$), possibly formed via the intermediacy of $CH_3\dot{N}_2$.

$$CH_3{-}N{=}N{-}CH_3 \xrightarrow{h\nu} \dot{C}H_3 + [\dot{N}_2CH_3] \rightarrow 2\dot{C}H_3 + N_2$$

$$\downarrow$$

$$C_2H_6$$

Irradiation of equimolar mixtures of azomethane and perdeuteroazomethane in the presence of oxygen as radical scavenger, produces nitrogen ($\varphi_{N_2} \simeq 0{\cdot}01$) and ethane containing no $CH_3CD_3$. This strongly suggests that the rearrangement

$$CH_3{-}N{=}N{-}CH_3 \xrightarrow{h\nu} C_2H_6 + N_2$$

is also likely as a subsidiary primary process.

In the liquid phase, the quantum yield is $< 0{\cdot}1$ and this may be interpreted either by a cage recombination, equation 4.2, or by a deactivation step, equation 4.3.

$$CH_3\dot{N}_2 + \dot{C}H_3 \rightarrow CH_3N_2CH_3 \tag{4.2}$$

$$CH_3N_2CH_3{}^{v} + CH_3N_2CH_3 \rightarrow 2CH_3N_2CH_3 \tag{4.3}$$

If cage recombination is an important process then the corollary is that the lifetime of $CH_3\dot{N}_2$ must be such as to allow reaction with $\dot{C}H_3$ within the solvent cage. Such a primary recombination may indeed take place in a relatively short time.

A common radical initiator containing the azo link is azobisisobutyronitrile, which on photolysis in benzene solution gives nitrogen with a quantum yield of 0·43 (cf. section 2.4).

At 313·0 nm the main process in diazirine is,[5]

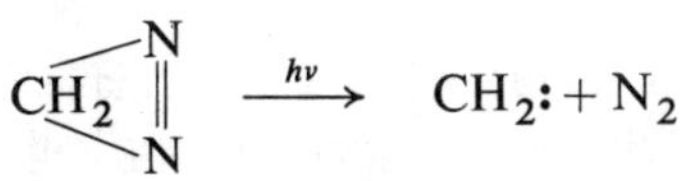

Photoisomerization to diazomethane also occurs. The photolysis of 3-methyldiazirine proceeds by two primary processes[6]. One gives rise to nitrogen and ethylene, equation 4.4, and the other to nitrogen and ethylidene, equation 4.5,

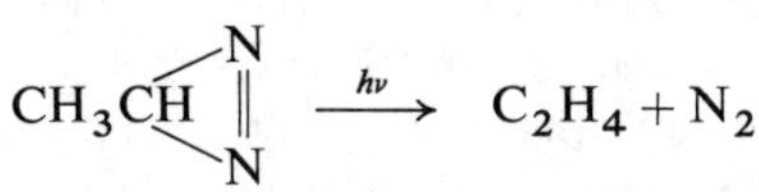

(4.4)

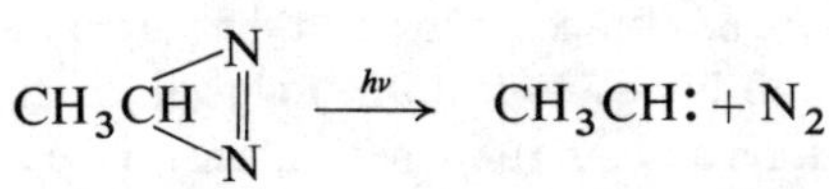

(4.5)

which undergoes the subsequent transformations (4.6), (4.7), and (4.8).

$$CH_3CH: \rightarrow C_2H_4{}^{v} \tag{4.6}$$

$$C_2H_4{}^{v} \rightarrow C_2H_2 + H_2 \tag{4.7}$$

$$C_2H_4{}^{v} \rightarrow C_2H_4 \tag{4.8}$$

Evidence is now available[7] concerning the geometry of nitrogen loss from cyclic azo compounds. Illumination of 3-carbomethoxy-*cis*-3,4-dimethyl-1-pyrazoline, **1**, results in stereospecific formation of 1-carbomethoxy-*cis*-1,2-dimethylcyclopropane, 2, whereas thermal decomposition proceeds with loss of stereochemistry. The *trans* isomer behaves correspondingly.

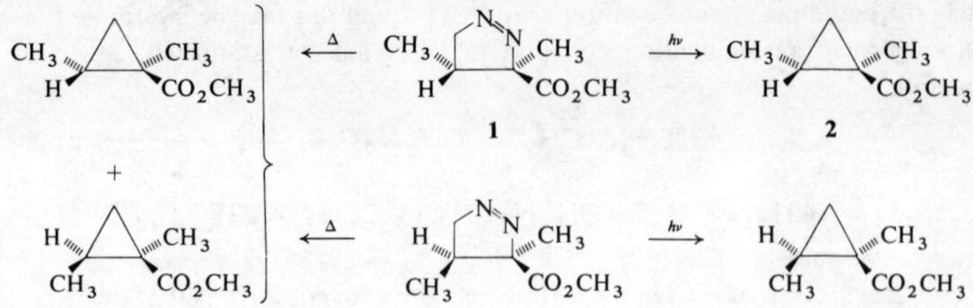

Irradiation of 4-methylene-$\Delta^1$-pyrazoline, **3**, at 315·0 nm yields[8] the theoretically important compound, trimethylenemethane, **4**.

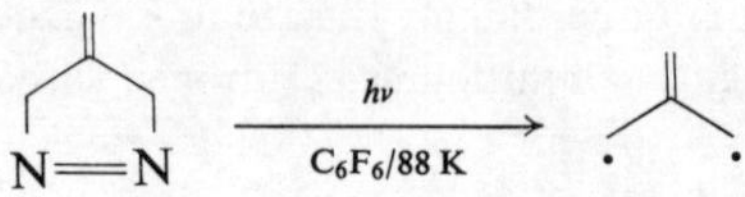

**3** **4**

The primary photodissociative process of diazomethane leads to methylene and nitrogen ($\varphi_{N_2} \simeq 4$), the following mechanism being indicated by flash photolysis studies:[9]

$$CH_2N_2 \xrightarrow{h\nu} CH_2N_2(S_1) \rightarrow CH_2{:}(S_1) + N_2 \xrightarrow[\text{gas}]{\text{inert}} CH_2{:}(T_0) + N_2$$

High inert gas pressures favour formation of triplet methylene and it was concluded that this is the ground state multiplicity. Chemical support has been gained[10] by photolysing diazomethane separately, in mixtures of varying composition of nitrogen and *trans*-but-2-ene, and nitrogen and *cis*-but-2-ene. When the partial pressure of nitrogen is low and the partial pressures of olefin and diazomethane are roughly equivalent, the following reactions, arising by operation of both addition and insertion mechanisms, and consistent with the above scheme, occur:

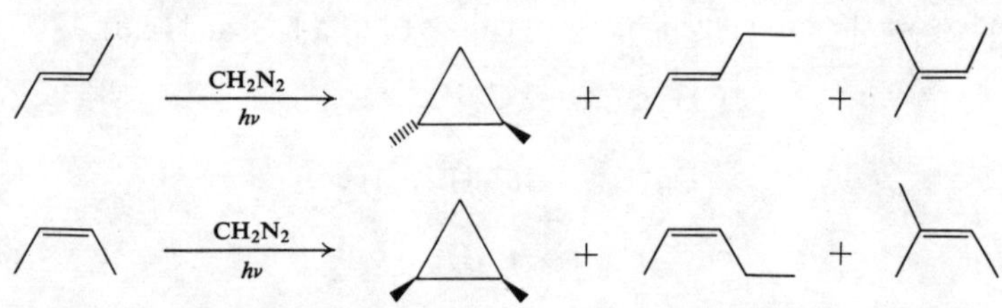

As the partial pressure of nitrogen increases and those of olefin and diazomethane drop, loss of stereospecificity is observed. Decomposition in the presence of benzophenone as photosensitizer also leads to non-stereospecific addition.

Irradiation of diazomethane in benzene has been claimed[11] to produce a mixture of toluene (by insertion), cycloheptatriene, **5**, and norcaradiene, **6**, though a convincing demonstration of the coexistence of these valence isomers is lacking.[12]

**5**

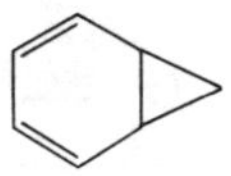

**6**

Diazoketones often undergo a rearrangement formally analogous to the Wolff rearrangement.

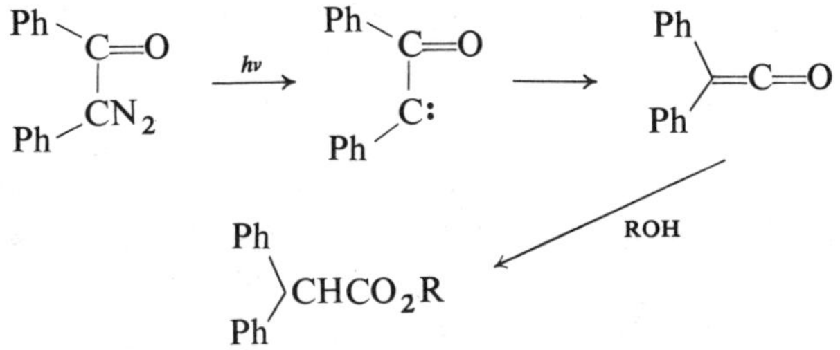

On irradiation, azides lose nitrogen and form nitrenes, **7**, which like methylene are believed to possess a triplet ground state.[13] Generally nitrenes react by rearrangement and abstraction mechanisms.[14, 15, 16]

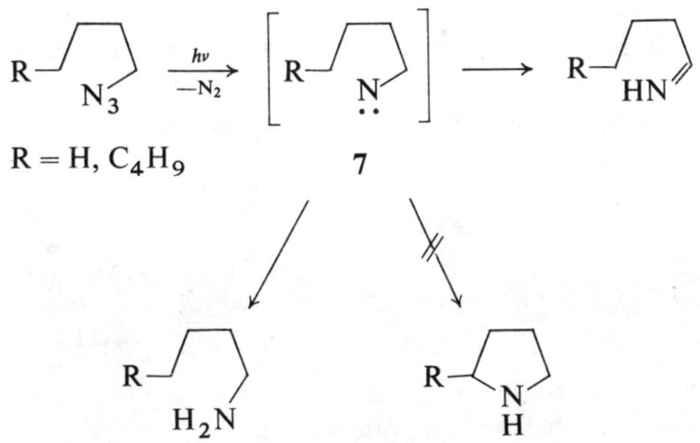

## *4.1.3 Elimination of nitric oxide*

Organic nitrites have a weak longest wavelength absorption corresponding to a $\pi^* \leftarrow n$ transition and involving a non-bonding nitrogen electron. Excitation $(\pi^* \leftarrow n)$ brings about photodissociation[17] of the RO–NO bond, thereby generating an alkoxy radical and nitric oxide:

$$\text{RO—NO} \rightarrow \text{R}\dot{\text{O}} + \text{N}\dot{\text{O}}$$

Alkoxy radicals may react in several ways.[18]

(a) Decomposition to carbonyl compounds, equation 4.9, or to alkyl radicals, equation 4.10.

$$\begin{matrix}R_1\\R_2\end{matrix}\!\!>\!CHONO \xrightarrow{h\nu} \begin{matrix}R_1\\R_2\end{matrix}\!\!>\!CH\dot{O} \longrightarrow \begin{matrix}R_1\\R_2\end{matrix}\!\!>\!C{=}O + H^{\bullet} \qquad (4.9)$$

$$\begin{matrix}CH_3\\CH_3(CH_2)_2\end{matrix}\!\!>\!C\!<\!\begin{matrix}CH_3\\ONO\end{matrix} \xrightarrow{h\nu} \begin{matrix}CH_3\\CH_3(CH_2)_2\end{matrix}\!\!>\!C\!<\!\begin{matrix}CH_3\\O^{\bullet}\end{matrix} \xrightarrow{NO}$$

$$(CH_3CH_2CH_2NO)_2 + (CH_3)_2CO \qquad (4.10)$$

C–C fission in the alkoxy radical, equation 4.10, is favoured over intramolecular hydrogen abstraction from a *primary* carbon and reactions of this sort may compete successfully with the Barton reaction (below). However, C–C fission is not observed when intramolecular hydrogen abstraction from a secondar or tertiary carbon is possible.

(b) Intermolecular hydrogen abstraction.

$$R\dot{O} + \text{Solv-H} \rightarrow ROH + \text{Solv}^{\bullet}$$

The extent to which intermolecular hydrogen abstraction occurs depends upon the efficiency of the solvent to function as a hydrogen donor.

(c) Intramolecular hydrogen abstraction. In certain instances, photolytically generated alkoxy radicals may react by stereoselective intramolecular hydrogen abstraction, followed by recombination of the resulting carbon radicals with nitric oxide, to form oximes or nitroso compounds. Such a sequence is known as the Barton Reaction.[19, 20]

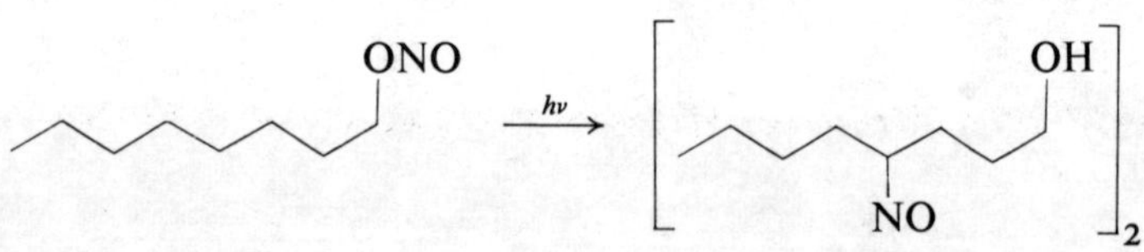


No evidence for substitution other than at $C_4$ has been observed, strongly suggesting a 6-membered cyclic transition state.

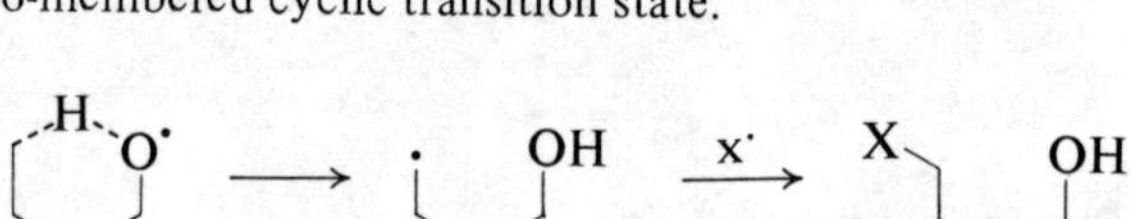


Support for this postulate has been obtained by Kabasakalian[18] who showed that despite the ease of benzyl hydrogen abstraction, 3-phenyl-1-propyl nitrite does not give rise to any nitroso dimer.

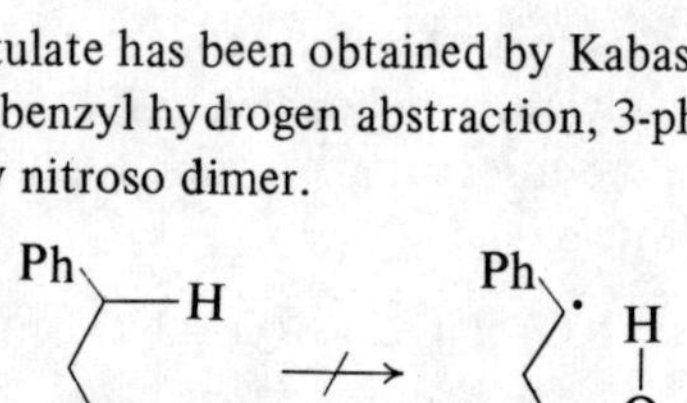

4-Phenyl-1-butyl nitrite, a compound which can form a 6-membered transition state, *does* react to give the corresponding nitroso dimer. The next higher homologue also reacts via a 6-membered transition state.

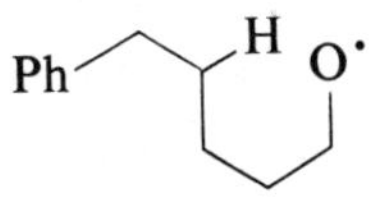


but not

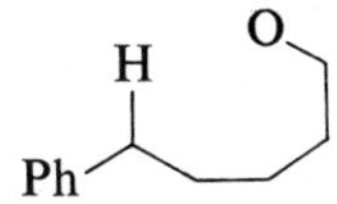


Barton found[21] this reaction to be of particular use in the steroid field for activating saturated carbon atoms.

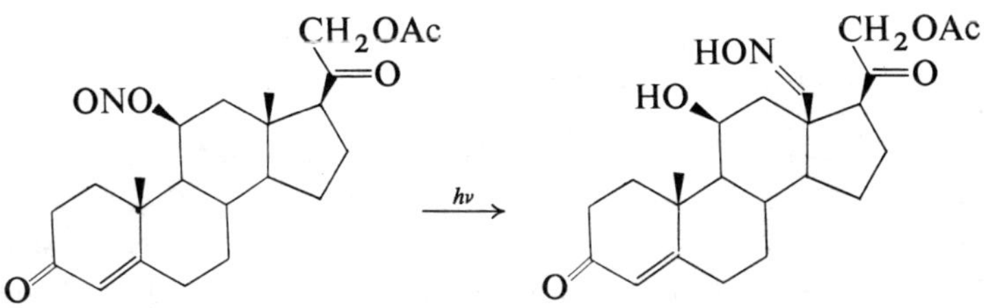


Corticosterone acetate nitrite

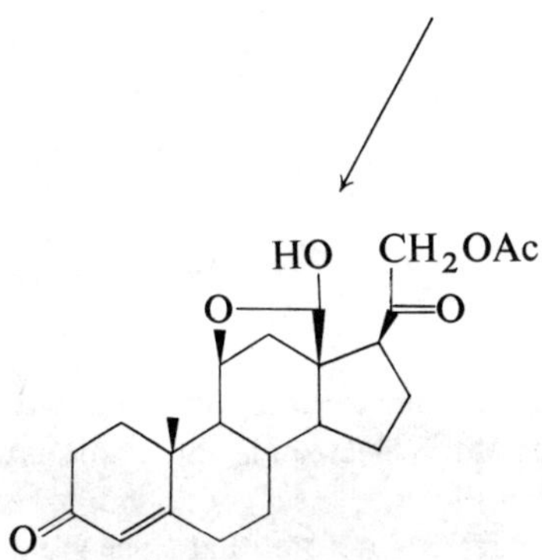


Aldosterone 21-acetate

### *4.1.4 Miscellaneous elimination reactions*

Irradiation of acyl hypoiodites formed either by the action of lead tetracetate/iodine or by *t*-butyl hypoiodite on the parent carboxylic acid provides an efficient method of photochemical decarboxylation and preparation of the corresponding nor-iodides.[22]

$$\mathrm{RCOOI} \xrightarrow{h\nu} \mathrm{RI} + \mathrm{CO_2}$$

Sulphones undergo photoelimination of sulphur dioxide[23] to yield a mixture of dienes. The stereochemistry of the major photoproduct is in accordance with a concerted conrotatory ring opening (cf. section 4.2.2).

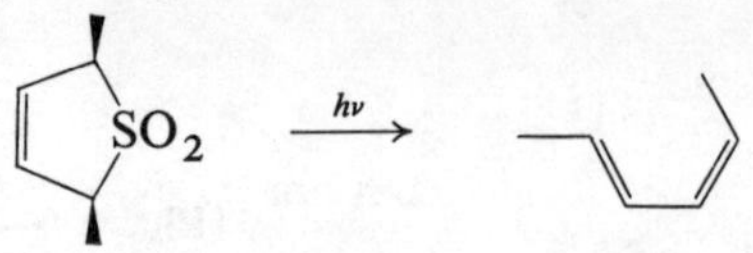

## 4.2 Rearrangements

### 4.2.1 *cis-trans Isomerization*

Calculations[24] show that ethylene singlets and triplets in their *lowest* vibrational state have a geometry very different from the ground state molecule. Instead of being planar with overlapping $p_z$ orbitals, the $sp^2$ hybridized carbon atoms have undergone a 90° dihedral twist to give a species in which the $p_z$ orbitals are mutually perpendicular (Fig. 4.2). This is because stabilization of the ground

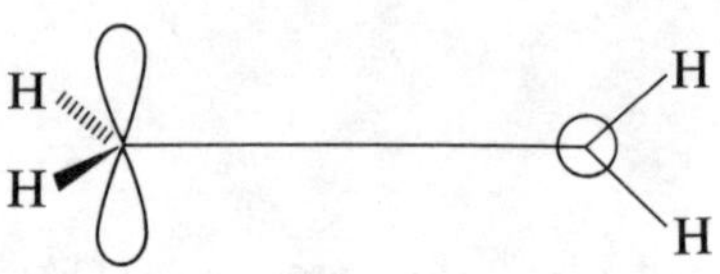

Fig. 4.2

state configuration by the one electron remaining in the $\pi$ MO is outweighed by the destabilizing effect of the electron in the $\pi^*$ MO. A potential energy diagram for ethylene is shown in Fig. 4.3.

It is reasonable to expect, therefore, that since vibrationally deactivated olefin singlets display such an orthogonal geometry, irradiation of olefins might bring about *cis-trans*-isomerization via the singlet state. The high energy necessary for the allowed $^1B_{1u} \leftarrow {}^1A_g$ transition ($\lambda_{max} \simeq 162$ nm) of simple olefins causes reactions other than isomerization, to be dominant. For example, *cis*-but-2-ene when irradiated (186–192 nm) suffers decomposition and forms high molecular weight products with some associated isomerization.[25]

Olefin triplets should similarly be capable of undergoing *cis-trans*-isomerization but in most simple olefins the difference in energy between the $S_1$ and $T_1$ levels large, and compared with other processes, intersystem crossing is inefficient.[25, 2] Olefin triplets can, however, be readily formed by use of a sensitizer and despite the fact that their degradation is fast, *cis* and *trans* geometric isomers can arise. The composition of the photostationary state is determined by several factors. Often the *trans* isomer absorbs more intensely at longer wavelengths than does the *cis* isomer. Assuming that the quantum yields for isomerization are similar, if long wavelength light is used for excitation, merely as a consequence of the differing extinction coefficients, the *cis* form will predominate. This process is an example of optical pumping (cf. section 5.1.2).

The photoisomerization of 1,2-dichloroethylenes has been studied[27] by irradiation under oxygen (130 bar pressure) whereby selective excitation to the triplet state takes place. Determination of the quantum yields for isomerization shows that although excitation of the two geometric isomers of 1,2-dichloroethylene leads initially to two discrete triplets, these decay into a common triplet faster than they can be quenched. Further, this common triplet decays to ground state *cis* and *trans* dichloroethylenes in the ratio 4:3. This same mechanism

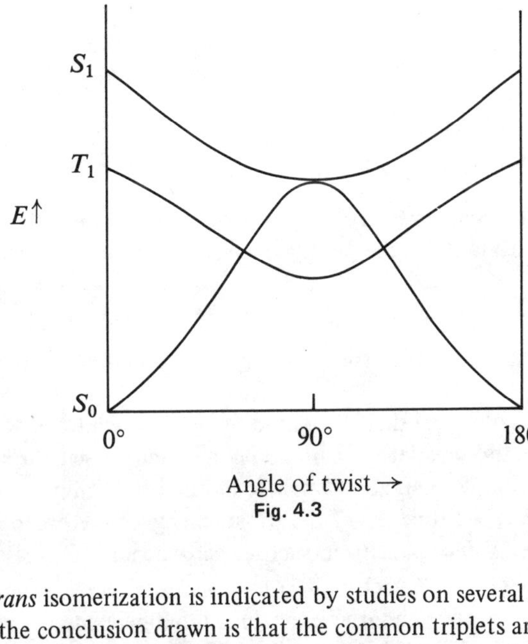


**Fig. 4.3**

for *cis* and *trans* isomerization is indicated by studies on several different compounds and the conclusion drawn is that the common triplets are lower in energy than those generated by vertical absorption of light and thus the lowest band measured for ethylene at 343 kJ is not a true 0–0 band. Instead it represents a transition to a highly vibrationally excited triplet. The molecule then undergoes vibrational cascade leading to the equilibrium conformation of the olefin triplet. It is well known that emission from olefin triplets cannot be detected, an observation now readily understood since radiative decay of a twist triplet to ground state is highly forbidden by the Franck–Condon principle. The foregoing results provide good experimental support for the theoretical predictions outlined earlier.

Just as ethylene may be regarded as a diradical in its first excited state, so may butadiene. Simple MO theory suggests a 1,4-diradical with severely restricted rotation about the $C_2$–$C_3$ bond, having the unpaired electrons at the termini of the system in orbitals which are mutually orthogonal, or nearly so.

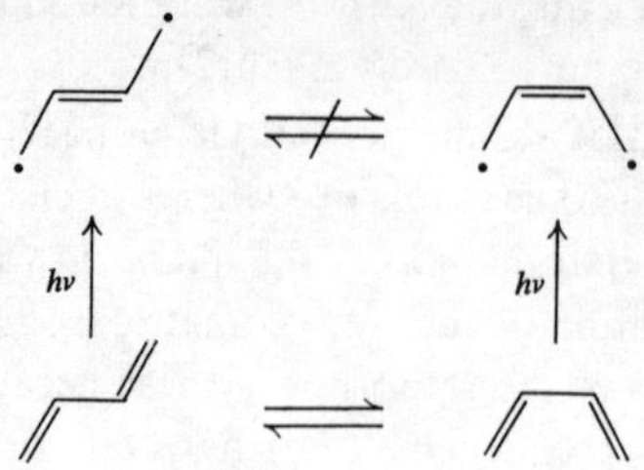


Trienes and tetraenes may be similarly considered as 1,6- and 1,8-diradicals, respectively.

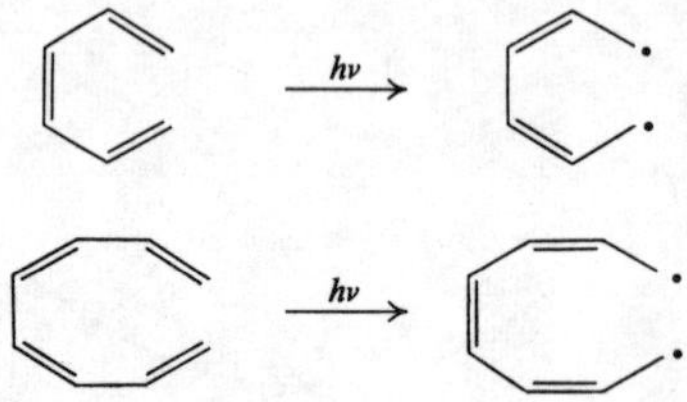


The benzene photosensitized geometric isomerization of *cis*- and *trans*-but-2-ene has been investigated[25] in the gas phase and takes place according to the mechanism

$$\text{Sens}\ (T_1) + \text{olefin}\ (S_0) \rightarrow \text{Sens}\ (S_0) + \text{olefin}\ (T_1)$$

It has been found that the olefin triplets decay to *cis* and *trans* isomers with equal probability.

Hammond[28] has observed that irradiation of benzene solutions of 1,3-pentadienes, containing a variety of carbonyl compounds as sensitizers, results in *cis-trans* isomerization. Remarkably, it was found that the composition of the photostationary state is a function of the triplet energy of the photosensitizer (Fig. 4.4). High energy photosensitizers such as acetophenone ($E_T = 310$ kJ mol$^{-1}$) and benzophenone ($E_T = 289$ kJ mol$^{-1}$) both gave the same composition of the photostationary state and it was concluded that here energy transfer to both the *trans* and *cis* triplets is efficient. As the sensitizer energy approached and finally fell below 247 kJ mol$^{-1}$, the composition of the photostationary state became *trans*-rich, indicating that the *trans* triplet has an energy of this order. Below 247 kJ mol$^{-1}$, the *trans* triplet is not efficiently populated. A maximum in the *trans*/*cis* ratio was reached about 238 kJ mol$^{-1}$ showing this to be the energy of the *cis* triplet. However, quenching of the sensitizer persisted down to 222 kJ mol$^{-1}$ and Hammond concluded that direct excitation of the twist triplet was occurring. Such triplets are often known as phantom or non-spectroscopic triplets since they are attained by a process, non-vertical transition, which is forbidden by the Franck–Condon principle. A scheme consistent with these observations is shown in Fig. 4.5.

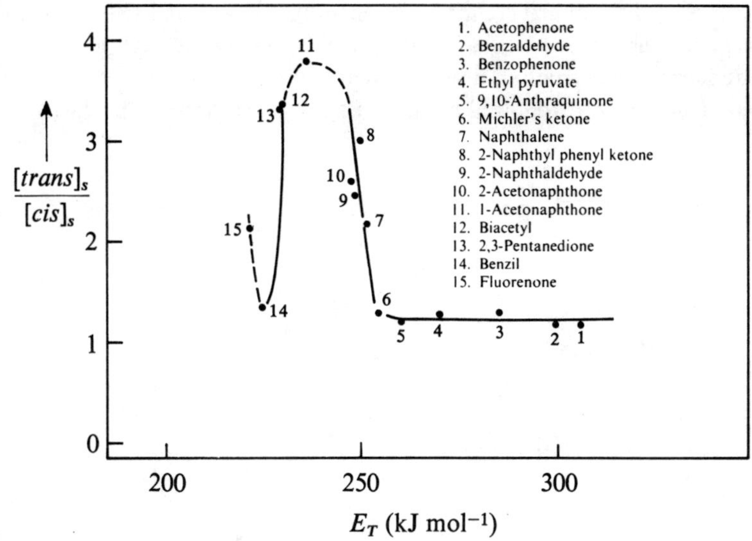


**Fig. 4.4** Sensitized isomerization of the piperylenes.

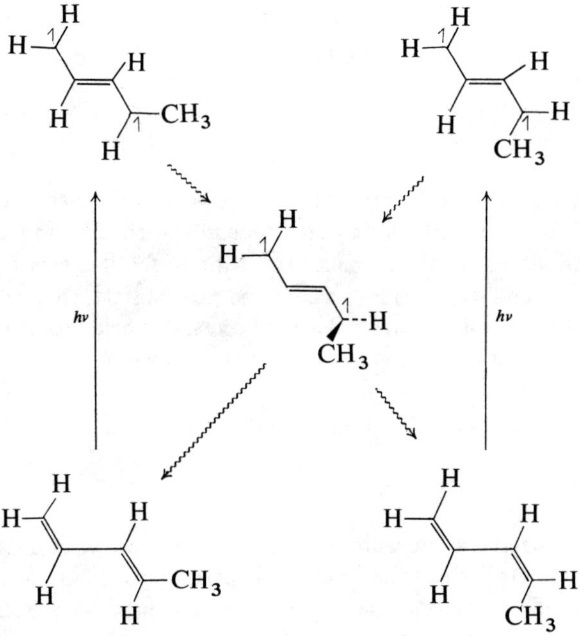

**Fig. 4.5**

Only one of the terminal groups in the phantom triplet is twisted, thereby allowing advantage to be taken of the resonance energy of the allyl group.

In stilbene[28] isomerization, three different triplets are also involved, the *cis* being of highest energy, followed by the twist and *trans* triplets (Fig. 4.6).

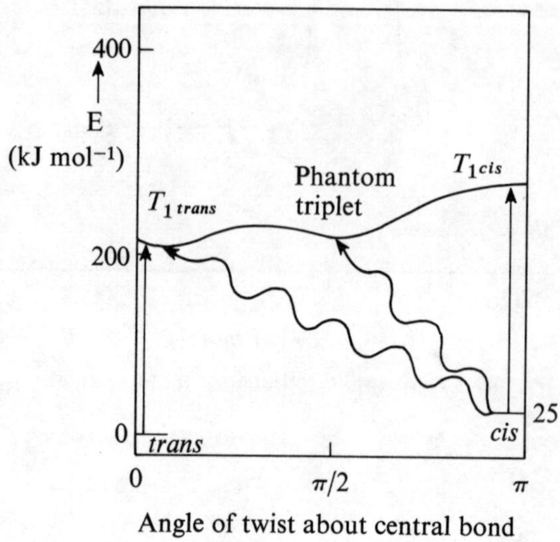


**Fig. 4.6** Potential function for rotation in stilbene triplet states.

*cis-trans* Isomerization is not confined to carbon systems. Azobenzene is known to undergo this reaction both directly and by sensitization. The results of the sensitized reactions lead to the conclusion[29] that transfer of energy to either stereoisomeric ground states leads to a common excited state, from which decay to both *cis* and *trans* ground state molecules occurs. It is believed that singlets generated in the direct irradiation also undergo isomerization.

## *4.2.2 Valence isomerization*

Valence isomerization is a reorganization of $\pi$, or of $\pi$ and $\sigma$ electrons accompani by changes in interatomic distances and bond angles but without migration of atoms or groups. Such reactions are widely known and have been extensively studied. For over half a century, investigations have been carried out[30] on the photochemical and thermal interconversions of ergosterol, **8**, lumisterol, **11**, precalciferol, **9**, calciferol, **10**, tachysterol, **12**, and suprasterol, **13**, and the following relationship established between them.

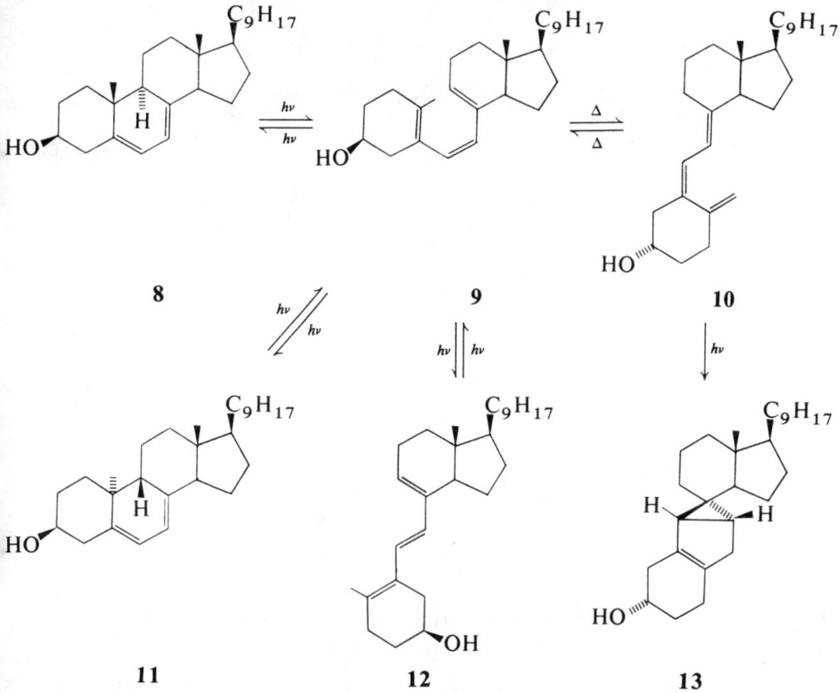


In its simplest form the ring fission of ergosterol is a transformation of the ype,

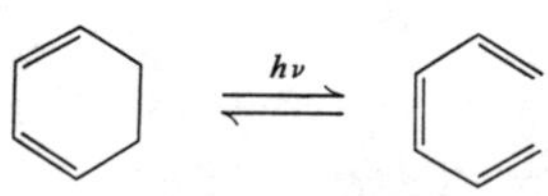


e., a general electrocyclic reaction,[1] the stereochemistry of which is governed y the principle of the conservation of orbital symmetry. Woodward and Hoffmann[1] ave discussed this reaction type in the ground and excited states, and a few xamples are given below:

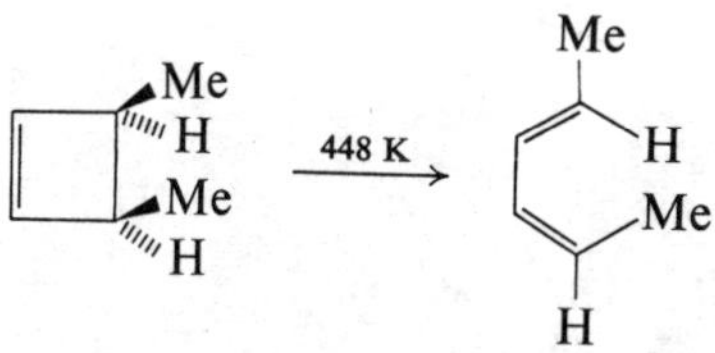


(Ref. 31)

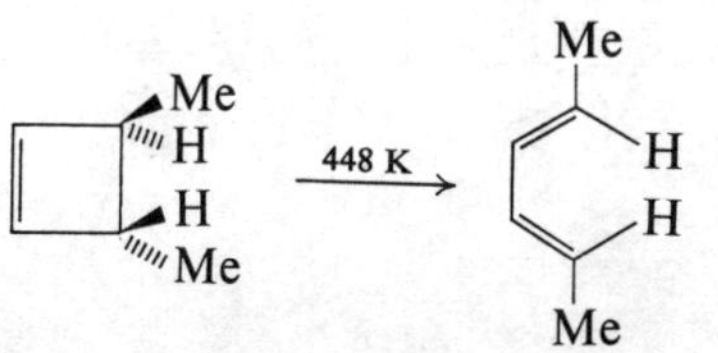


(Ref. 31)

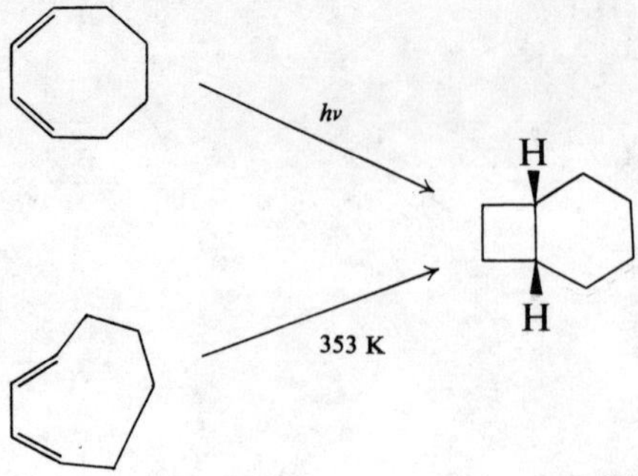


(Ref. 32)

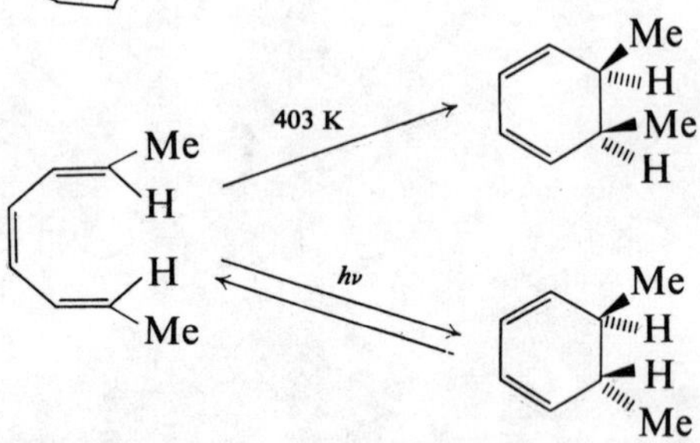


(Ref. 33)

(Ref. 34)

That these stereochemistries are a direct consequence of the symmetries of the orbitals involved in the reaction, is demonstrated here in two ways.

(a) There are two equivalent representations of the Hückel molecular orbitals (HMO's) of butadiene (Fig. 4.7) and it is the symmetry of the highest occupied molecular orbital (HOMO) which determines the stereochemistry of the reaction

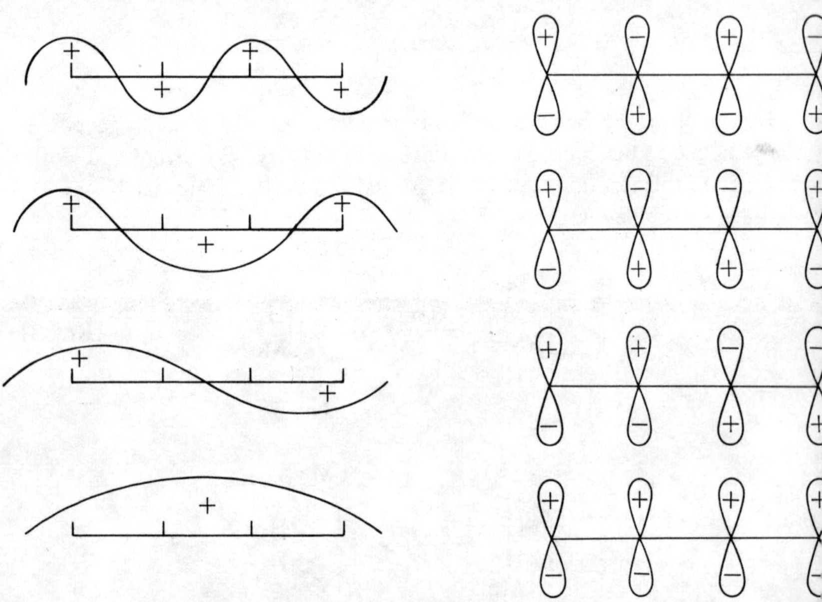

Fig. 4.7

The transformation of butadiene to cyclobutene requires interaction of the $p$-orbitals situated on $C_1$ and $C_4$ and such interaction requires a formal rotation of these $p$-orbitals, which, *a priori,* can take place in two ways, designated, respectively, conrotatory and disrotatory (Fig. 4.8).

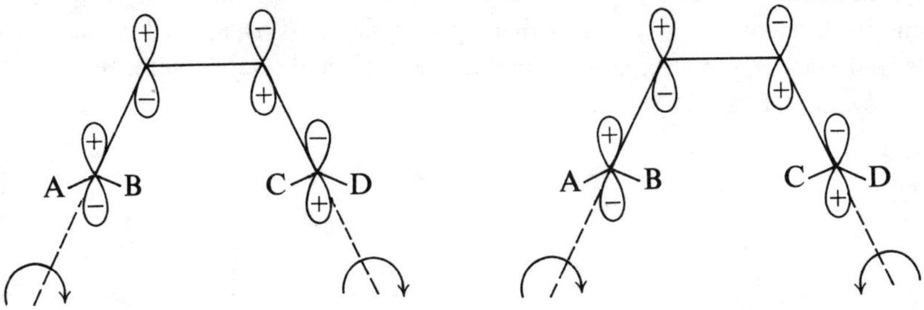


Conrotation

Disrotation

**Fig. 4.8**

Conrotation leads to both positive lobes (or both negative lobes if rotation of the $p$-orbitals is anticlockwise) appearing in the region between $C_1$ and $C_4$ and gives rise to a bonding situation. Conversely, disrotation leads to an anti-bonding situation. The result follows, that since only one of two possible rotation modes leads to accumulation of electron density between $C_1$ and $C_4$ and hence to reaction, then the stereochemistry of the product is uniquely determined by the stereochemistry of the starting material.

In the case of a photoexcited molecule, the HOMO is the singly occupied $\psi_3$ (Fig. 4.9), and it is the disrotatory mode which gives rise to ring closure. It is to

Conrotation

Disrotation

**Fig. 4.9**

be noticed, however, that the stereochemistry is precisely opposite to that observed in the thermal reaction.

This same procedure can be applied to any electrocyclic reaction irrespective of the number of $\pi$ electrons involved and it leads to the conclusion that for systems possessing $4n + 2$ $\pi$-electrons, disrotatory ring closure is thermally allowed and conrotatory ring closure is photochemically allowed, whereas for systems possessing $4n$ $\pi$-electrons, conrotatory closure is thermally allowed and disrotatory closure is photochemically allowed.

(b) An alternative but more rigorous way of arriving at the same conclusion is to classify the reactant and product orbitals according to the symmetry element preserved along the reaction co-ordinate. For example, the conrotatory ring opening of cyclobutene to give butadiene preserves a $C_2$-axis, located in the molecular plane and bisecting the cyclobutene single and double bonds involved in the reaction (Fig. 4.10).

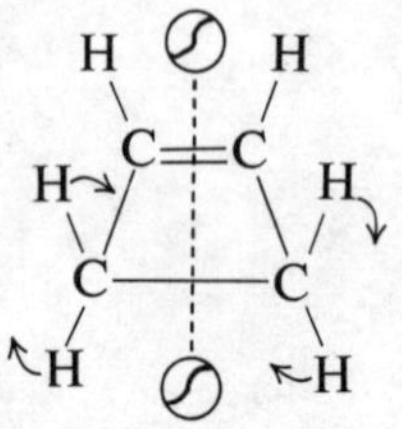

**Fig. 4.10**

With respect to this symmetry element, the $\sigma$-bond is symmetric (S), and, conversely, the $\pi$-bond is antisymmetric (A). Choosing the same axis and classifying the butadiene product orbitals with respect to it, leads to the assignments shown in Fig. 4.11.

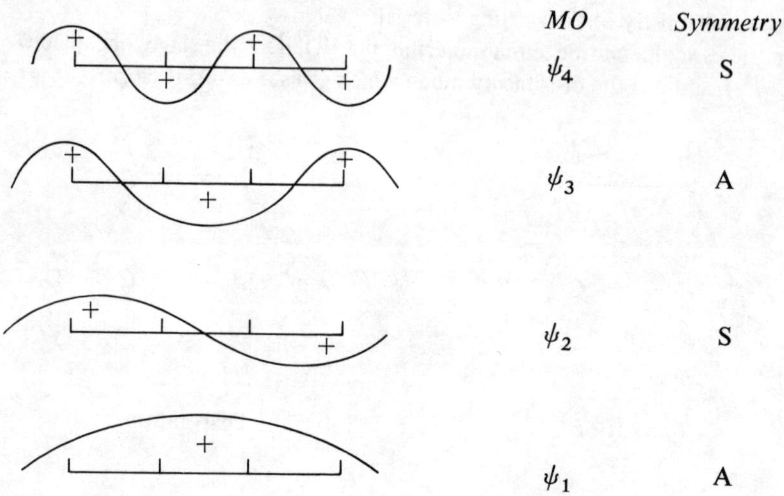


**Fig. 4.11**

A correlation diagram can now be constructed by joining reactant and product levels of like symmetry, from which it is clear, that if the symmetry of the orbitals involved in the reaction is preserved, then the orbitals of ground state cyclobutene correlate with the orbitals of ground state butadiene, there will be no symmetry imposed barrier to the transformation, and the reaction is thermally allowed (Fig. 4.12).

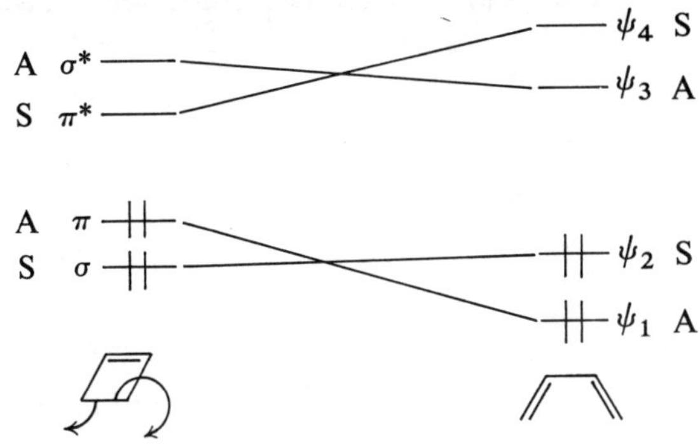


Fig. 4.12

The alternative construction that can be arrived at by joining levels of like symmetry, i.e., by joining $\sigma$ and $\psi_4$, etc., breaks the quantum mechanical non-crossing rule[35] which requires that only levels of unlike symmetry may cross.

If, on the other hand, the corresponding diagram is constructed for the conversion of electronically excited cyclobutene to butadiene, it is found that the orbitals of the cyclobutene correlate, not with those of excited butadiene, in which $\psi_1$ is doubly occupied and $\psi_2$ and $\psi_3$ are both singly occupied, but with a high energy form in which $\psi_2$ is doubly occupied and $\psi_1$ and $\psi_4$ are singly occupied. The photochemically induced conrotatory ring opening of cyclobutene to butadiene is, therefore, symmetry forbidden.

The disrotatory ring opening of cyclobutene to butadiene preserves a mirror plane situated at right angles to the molecular plane and bisecting the $\pi$ bond (Fig. 4.13).

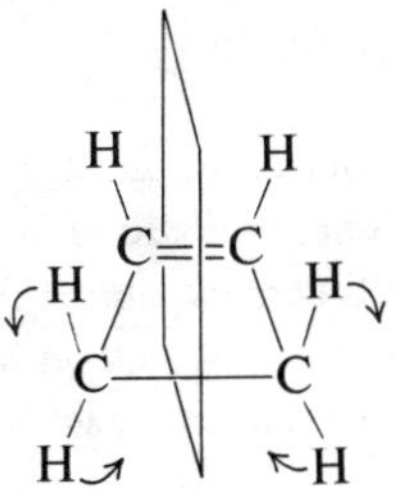

Fig. 4.13

The cyclobutene $\sigma$, $\pi$, $\pi^*$, and $\sigma$ MO's show S, S, A, and A symmetries respectively with respect to this plane and the butadiene $\psi_1$, $\psi_2$, $\psi_3$, and $\psi_4$ MO's show S, A, S, and A symmetries, respectively. The only possible correlation diagram which observes the quantum mechanical non-crossing rule is shown in Fig. 4.14, from which it can be seen that ground state cyclobutene correlates with a doubly excited form of butadiene, rendering the transformation thermally forbidden. Excited cyclobutene, however, correlates with excited butadiene and the photochemical transformation is consequently symmetry allowed (Fig. 4.15).

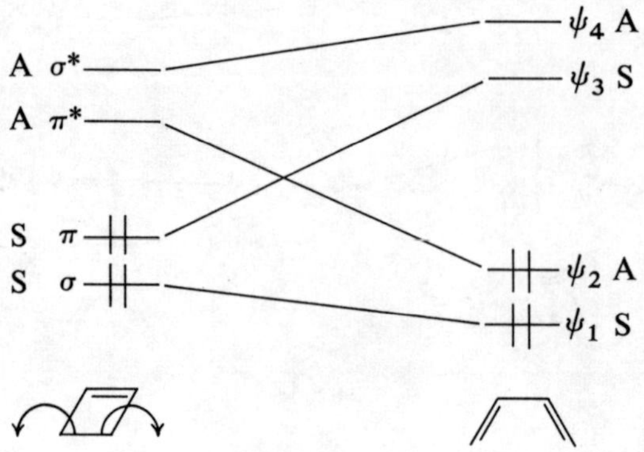


Fig. 4.14

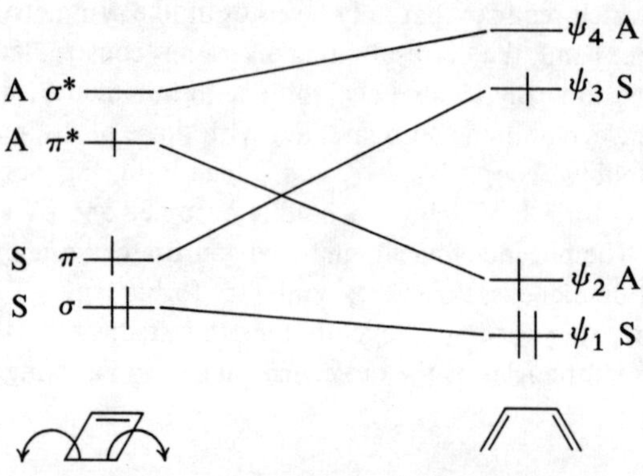


Fig. 4.15

It must be stressed, however, that although a particular reaction may be symmetry allowed, for thermodynamic reasons, it may still not occur.

Similar analyses demonstrate that these stereochemical conclusions apply to all electrocyclic reactions involving $4n$ $\pi$-electrons and it can also be shown that precisely opposite consequences obtain for systems involving $4n + 2$ $\pi$-electrons, as summarized in Table 4.1.

TABLE 4.1: Allowed modes of ring closure in electrocyclic reactions

| Number of electrons | Ground state | Excited state |
|---|---|---|
| $4n$ | Conrotatory | Disrotatory |
| $4n + 2$ | Disrotatory | Conrotatory |

These results are amply borne out by experiment. For example, Marvel and Seubert[36] have performed the transformations illustrated below and have found the stereochemistry as indicated:

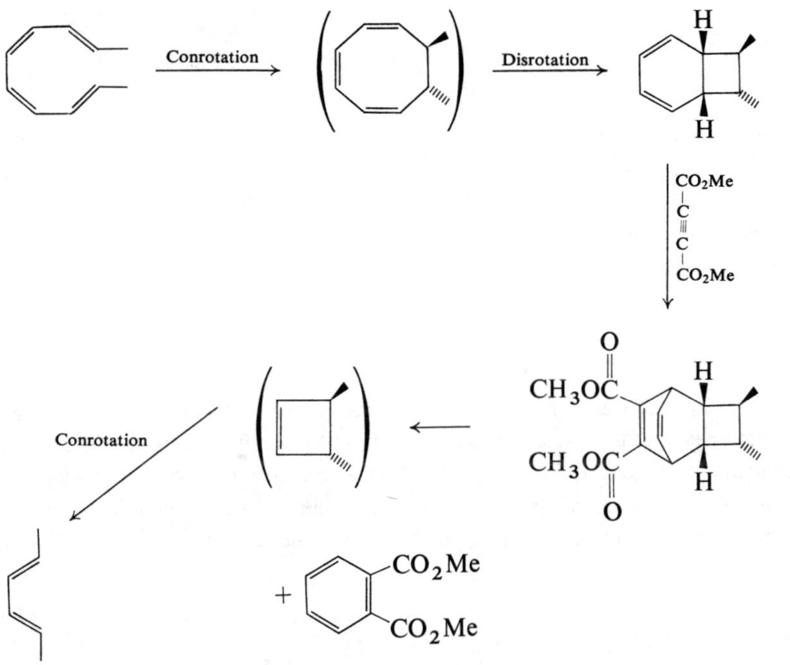


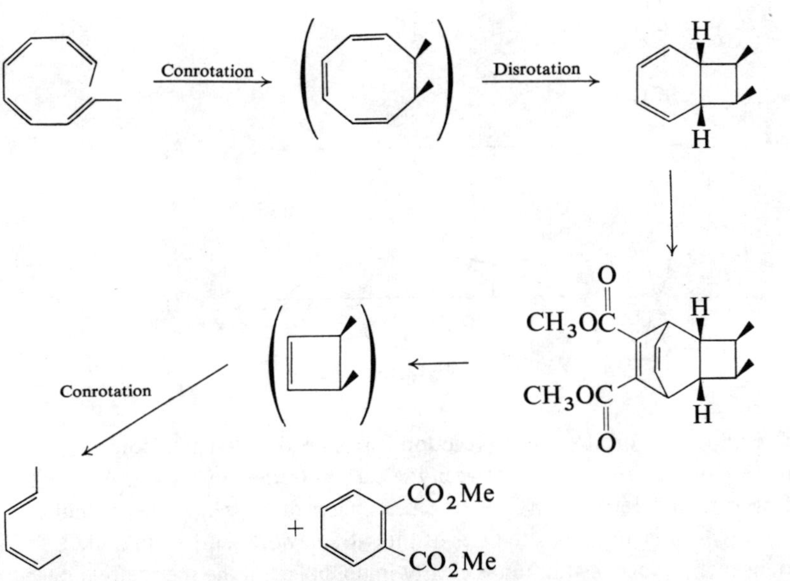

It is convenient at this point to consider the closely related *sigmatropic* rearrangements, which are defined as the migration of a $\sigma$-bond linked to one or more $\pi$-systems to a new position within that system. A typical example is the thermally induced Cope rearrangement,

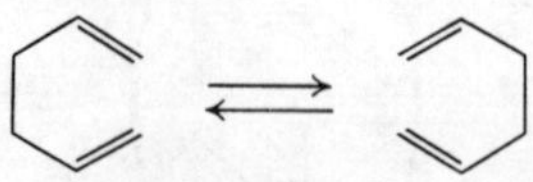

in which each terminus of the migrating $\sigma$-bond undergoes a [1, 3] shift. The reaction is consequently known as a [3, 3] sigmatropic rearrangement. Generally, if the termini move through *i*-1 and *j*-1 atoms respectively, then the transformation is classified as a sigmatropic rearrangement of order [*i, j*].

Consider the [1, 3] sigmatropic rearrangement.

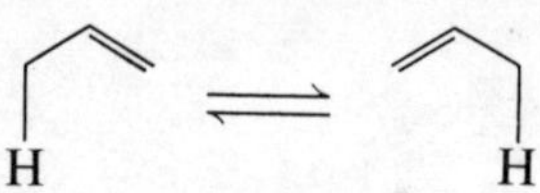


Since neither the starting material nor product contains a significant molecular symmetry element, correlation diagrams are not relevant despite the fact that the transition state, which, if it is regarded formally as being derived from a hydrogen atom and an allyl radical, possesses a mirror plane. Two equivalent ways of representing the MO's of the allyl radical are shown in Fig. 4.16.

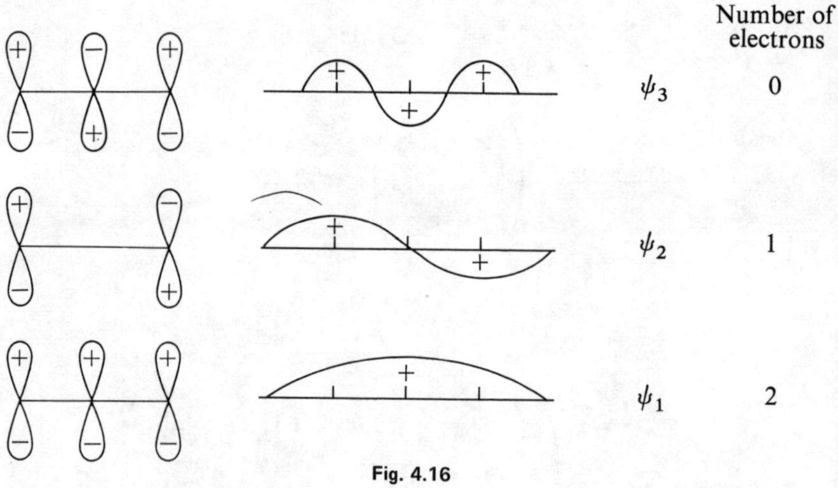


Fig. 4.16

The important orbital for the reaction is $\psi_2$, i.e., the HOMO. However, formation of a 3-centre bond between the carbon termini of $\psi_2$ and the 1*s* hydrogen orbital demands overlap between phases of $\psi_2$ which are of similar sign, but which lie on opposite faces of the $\pi$-framework, i.e., an antarafacial rearrangement. Such a situation is clearly impossible and the thermally induced

[1, 3] sigmatropic rearrangement is symmetry forbidden. However, in the corresponding photochemical reaction, the HOMO is $\psi_3$, a 3-centre bond can be formed by a suprafacial bridging of the 1,3-carbon atoms of the allyl radical, and the reaction is symmetry allowed. An identical result will be found for all sigmatropic rearrangements of this type in which $1 + j = 4n$.

By contrast, the [1, 5] sigmatropic rearrangement and, indeed, all rearrangements of this type in which $1 + j = 4n + 2$, can, by similar analysis, be shown to be thermally allowed and photochemically forbidden. These conclusions are shown in Table 4.2.

TABLE 4.2: Selection rules for sigmatropic reactions of order $[i, j]$ with $i = 1$ and $j > 1$

| $[i + j]$ | Ground state | Excited state |
|---|---|---|
| $4n$ | antarafacial | suprafacial |
| $4n + 2$ | suprafacial | antarafacial |

In the foregoing cases, interaction in the transition state has been allowed between a σ-orbital of the migrating group and the π-system, and reaction has taken place with retention of configuration at the migrating centre. When the migrating group possesses an accessible π-orbital, a set of selection rules precisely opposite to those in Table 4.2 apply and inversion of the migrating centre occurs.

If both $i$ and $j > 1$, topological distinctions for both π-systems must be made and the selection rules for these cases are summarized in Table 4.3.

TABLE 4.3: Selection rules for sigmatropic reactions of order $[i, j]$ with $i$ and $j > 1$

| $i + j$ | Ground state | Excited state |
|---|---|---|
| $4n$ | antara-supra<br>supra-antara | supra-supra<br>antara-antara |
| $4n + 2$ | supra-supra<br>antara-antara | antara-supra<br>supra-antara |

Two examples of sigmatropic shifts are given.

(a) A photochemical [1, 3] shift.[37]

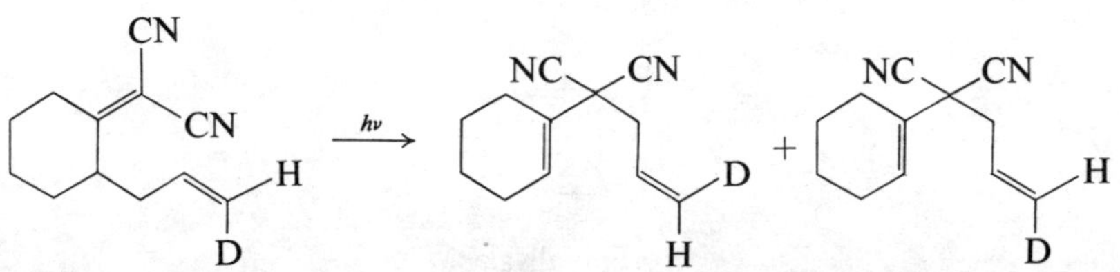

(b) A thermal [1, 3] shift proceeding with inversion at the migrating centre.[38]

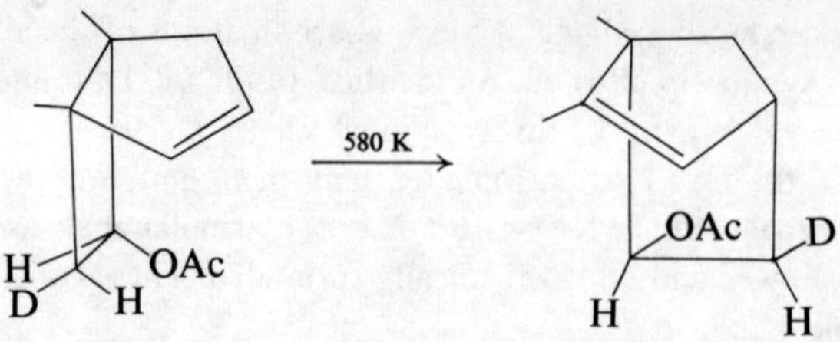


Valence isomerizations of non-conjugated olefins are also known. For example, the ring closure of norbornadiene to quadricyclane may be induced by direct or by sensitized[39] irradiation.

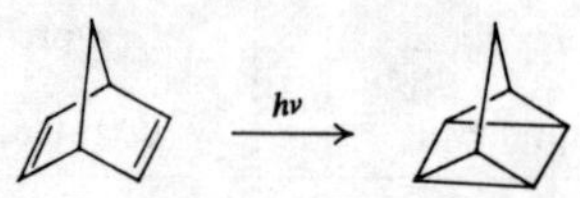


Zimmerman has investigated[40] the acetone-sensitized conversion of barrelene to semibullvalene, an example of the general[41] transformation divinylmethane → vinylcyclopropane, and has shown it to proceed by the mechanism:

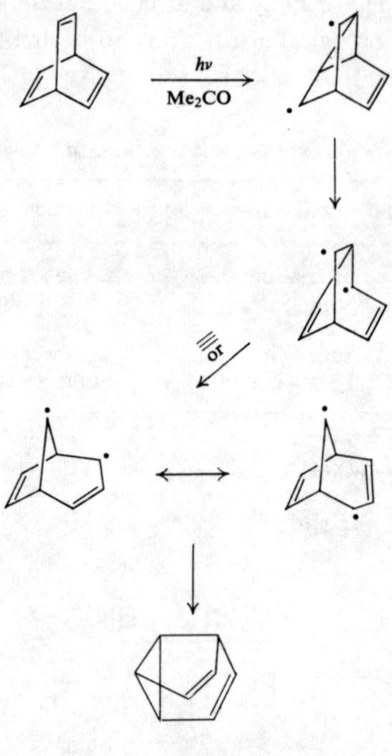


Semibullvalene

Photoisomerizations of aromatic rings have lead to some interesting photochemistry and benzene, in particular, has been the subject of extensive study.[42] The states of benzene in order of increasing energy are shown[43] in Fig. 4.17,

$$\tilde{X}(^1A_{1g}):\ a_{2u}^{\ 2}\, e_{1g}^{\ 4}$$
$$\tilde{A}(^1B_{2u}):\ a_{2u}^{\ 2}\, e_{1g}^{\ 3}\, e_{2u}$$
$$\tilde{B}(^1B_{1u}):\ a_{2u}^{\ 2}\, e_{1g}^{\ 3}\, e_{2u}$$
$$\tilde{C}(^1E_{2g}):\ a_{2u}^{\ 2}\, e_{1g}^{\ 2}\, e_{2u}^{\ 2}$$
$$\tilde{D}(^1E_{1u}):\ a_{2u}^{\ 2}\, e_{1g}^{\ 3}\, e_{2u}$$

**Fig. 4.17**

and give rise to the symmetry forbidden transitions, $^1B_{2u} \leftarrow {}^1A_{1g}$ ($\lambda_{max} \simeq 256$ nm) and $^1B_{1u} \leftarrow {}^1A_{1g}$ ($\lambda_{max} \simeq 200$ nm) and the symmetry allowed transition $^1E_{1u} \leftarrow {}^1A_{1g}$ ($\lambda_{max} \simeq 180$ nm).

The principal photochemical reactions of benzene can be summarized as follows:

(a) Irradiation of benzene in the liquid phase produces fulvene and benzvalene.[44]

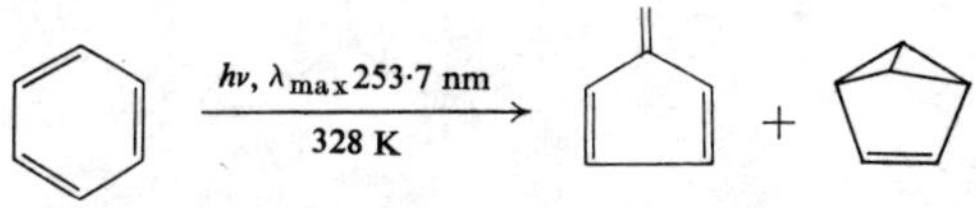


Increases in temperature promote formation of the former isomer, indicating that both electronic and vibrational excitation are necessary. This is supported by the observation that irradiation of benzene vapour at 184·9 nm also gives fulvene.[45]

(b) 1,2,4- and 1,3,5-Tri-*t*-butylbenzene as well as hexafluorobenzene give mixtures of the corresponding Dewar benzene (bicyclohexadiene), benzvalene, and prismane isomers.[46]

(c) Liquid phase irradiation of benzene-olefin mixtures gives one kind of adduct only,[47,48] structures of the type **14** and **15** not being found (cf. section 4.3.9),

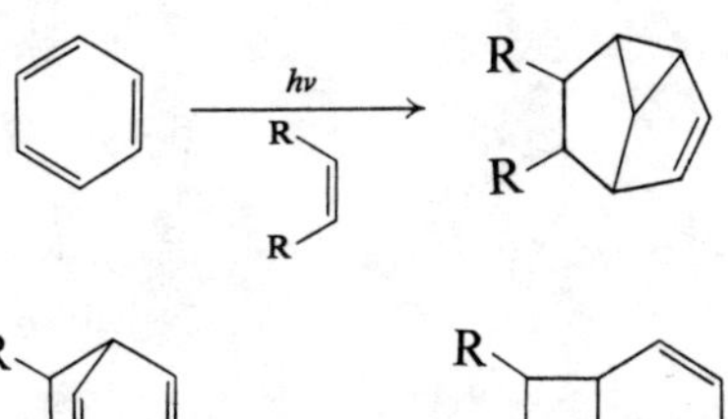


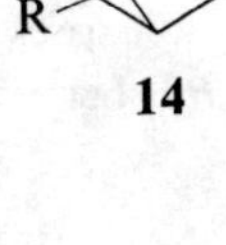


**14**

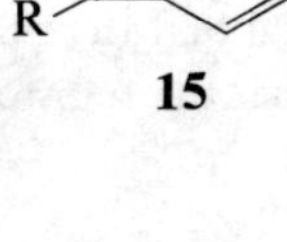


**15**

(d) 1,4-Addition across the *para*-positions occurs with conjugated dienes,[49] e.g.,

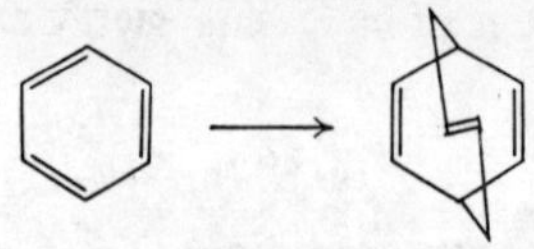

Liquid phase irradiation of benzene at 253·7 nm populates the $^1B_{2u}$ state wh under a distortion in which $C_1$ of the ring is displaced in the positive $z$ direction, may valence tautomerize[50] to prefulvene via a transition state of point group $C_s$ (Fig. 4.18).

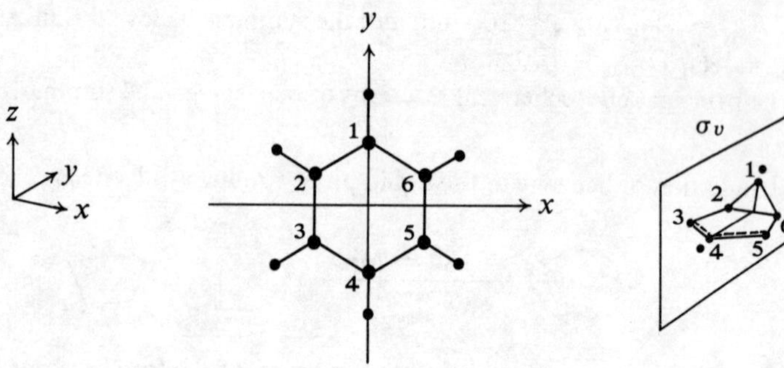


Fig. 4.18

The prefulvene symmetrized MO's and electronic configurations of the variou states together with the orbital and state correlation diagrams are shown in Fig. 4.19, from which it is also clear that the $^1B_{1u}$ state may not undergo a similar transformation. That distortion is necessary accounts for the need for vibrationa excitation. Reaction of prefulvene with olefin will lead to the observed product

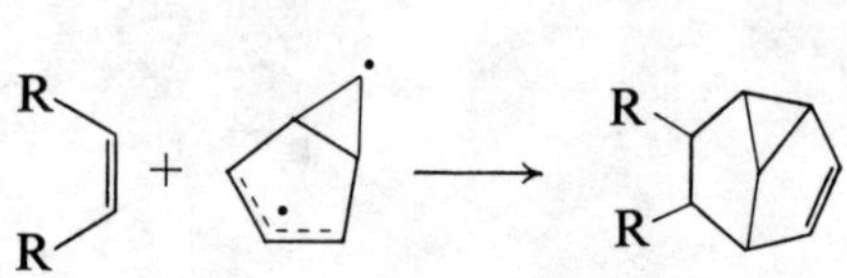


**16**

It has been suggested that the benzvalene skeleton arises from prefulvene[42] and it can be shown that the transformation of $^1B_{1u}$ benzene to benzvalene is

$\tilde{X}(A''): \psi_1^2 \psi_2^2 \psi_3 \psi_4$
$\tilde{A}(A'): \psi_1^2 \psi_2^2 \psi_3^2$
$\tilde{B}(A'): \psi_1^2 \psi_2^2 \psi_4^2$
$\tilde{C}(A'): \psi_1^2 \psi_2 \psi_3^2 \psi_4$
$\tilde{D}(A''): \psi_1^2 \psi_2 \psi_3 \psi_4^2$
$\tilde{E}(A''): \psi_1^2 \psi_2^2 \psi_3 \psi_5$
$\tilde{F}(A'): \psi_1^2 \psi_2^2 \psi_4 \psi_5$

(a)

| | |
|---|---|
| $\psi_1 = \sigma_{26}$ | $a'$ |
| $\psi_2 = p_3 + \sqrt{2}p_4 + p_5$ | $a'$ |
| $\psi_3 = p_3 - p_5$ | $a''$ |
| $\psi_4 = p_1$ | $a'$ |
| $\psi_5 = p_3 - \sqrt{2}p_4 + p_5$ | $a'$ |
| $\psi_6 = \sigma_{26}^*$ | $a''$ |

(b)

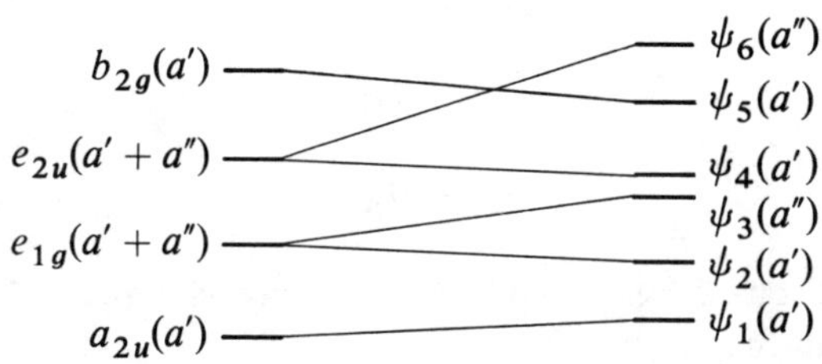


(c)

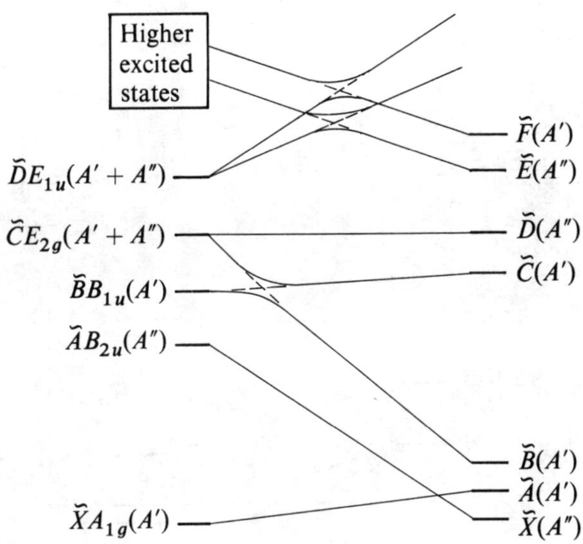


(d)

**Fig. 4.19**
(a) Electronic configurations of the states of prefulvene.
(b) Symmetrized MO's of prefulvene. For antibonding orbitals, the positive lobe is taken to be near the C atom of lower number.
(c) Orbital correlation diagram for the transformation, benzene → prefulvene.
(d) State correlation diagram for the transformation, benzene → prefulvene.

symmetry allowed,[50] the transition state belonging to the $C_2$ point group, in that $C_5$ and $C_6$ of the benzene ring are displaced in the positive and negative $z$ directions respectively (Fig. 4.20).

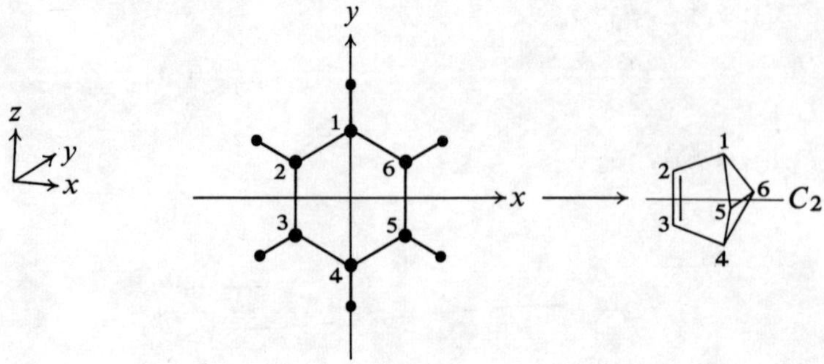


Fig. 4.20

The orbital and state correlation diagrams together with other relevant information are shown in Fig. 4.21.

$\tilde{X}(A)$: $\psi_1^2\,\psi_2^2\,\psi_3^2$
$\tilde{A}(B)$: $\psi_1^2\,\psi_2^2\,\psi_3\,\psi_4$
$\tilde{B}(A)$: $\psi_1^2\,\psi_2\,\psi_3^3\,\psi_4$
$\tilde{C}(B)$: $\psi_1\,\psi_2^2\,\psi_3^2\,\psi_4$
$\tilde{D}(A)$: $\psi_1^2\,\psi_2^2\,\psi_4^2$

| | |
|---|---|
| $\psi_1 = \sigma_{15} + \sigma_{46}$ | $a$ |
| $\psi_2 = \sigma_{15} - \sigma_{46}$ | $b$ |
| $\psi_3 = \pi_{23}$ | $a$ |
| $\psi_4 = \pi^*_{23}$ | $b$ |
| $\psi_5 = \sigma^*_{15} - \sigma^*_{46}$ | $b$ |
| $\psi_6 = \sigma^*_{15} + \sigma^*_{46}$ | $a$ |

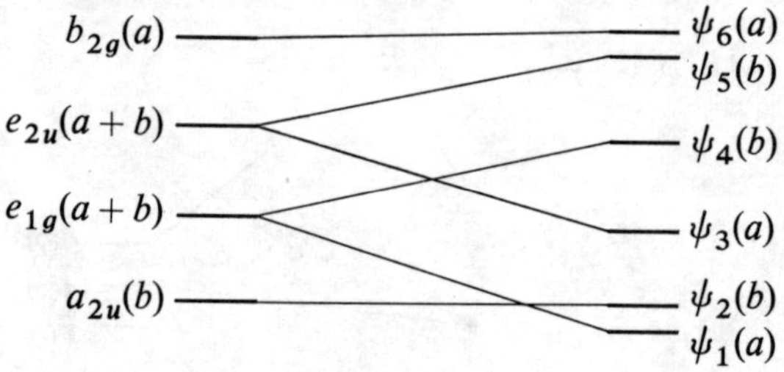


Fig. 4.21
(a) Electronic configurations of the states of benzvalene.
(b) Symmetrized MO's of benzvalene. For antibonding orbitals, the positive lobe (in the positive $z$ direction for $\pi^*$ and along the bond axis for $\sigma^*$) is taken to be near the C atom of lower number.
(c) Orbital correlation diagram for the transformation, benzene → benzvalene.

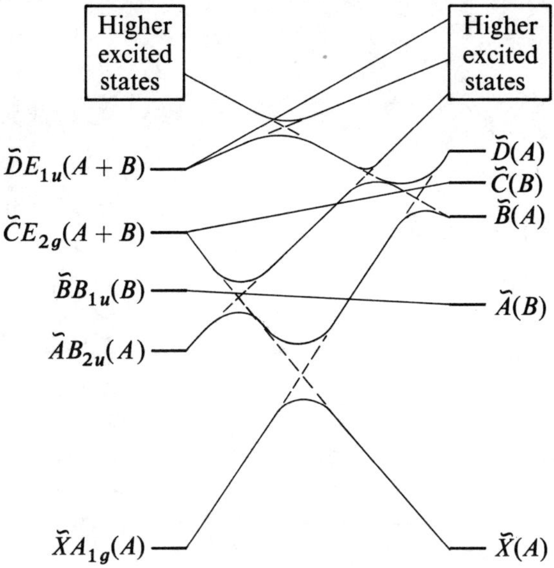


**Fig. 4.21**—*continued*
(d) State correlation diagram for the transformation, benzene → benzvalene.

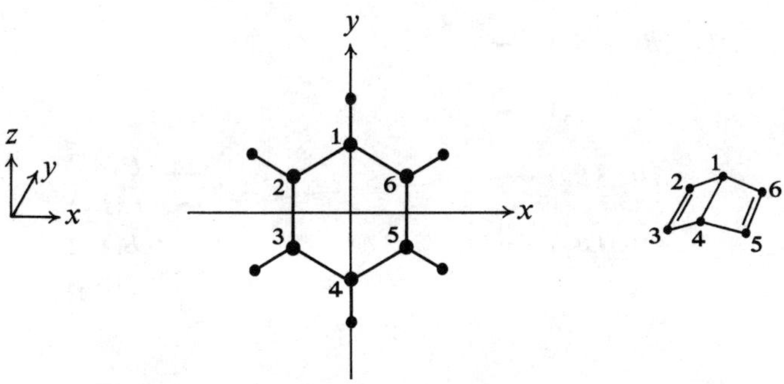


**Fig. 4.22**

Intersystem crossing leads to the $^3B_{1u}$ state which distorts spontaneously[51] and which may proceed to an excited form of Dewar benzene via a transition state of point group $C_{2v}$, in which $C_1$ and $C_4$ have been displaced in the positive $z$ direction (Fig. 4.22). The orbital and state correlation diagrams and other relevant information are shown in Fig. 4.23:

$\tilde{X}(A_1)$: $\psi_1^2 \psi_2^2 \psi_3^2$

$\tilde{A}(A_2)$: $\psi_1^2 \psi_2^2 \psi_3 \psi_4$

$\tilde{B}(B_2)$: $\psi_1^2 \psi_2^2 \psi_3 \psi_5$

$\tilde{C}(B_2)$: $\psi_1^2 \psi_2 \psi_3^2 \psi_4$

$\tilde{D}(A_2)$: $\psi_1^2 \psi_2 \psi_3^2 \psi_5$

(a)

| | |
|---|---|
| $\psi_1 = \sigma_{14}$ | $a_1$ |
| $\psi_2 = \pi_{23} + \pi_{56}$ | $a_1$ |
| $\psi_3 = \pi_{23} - \pi_{56}$ | $b_1$ |
| $\psi_4 = \pi^*_{23} - \pi^*_{56}$ | $b_2$ |
| $\psi_5 = \pi^*_{23} + \pi^*_{56}$ | $a_2$ |
| $\psi_6 = \sigma^*_{14}$ | $b_2$ |

(b)

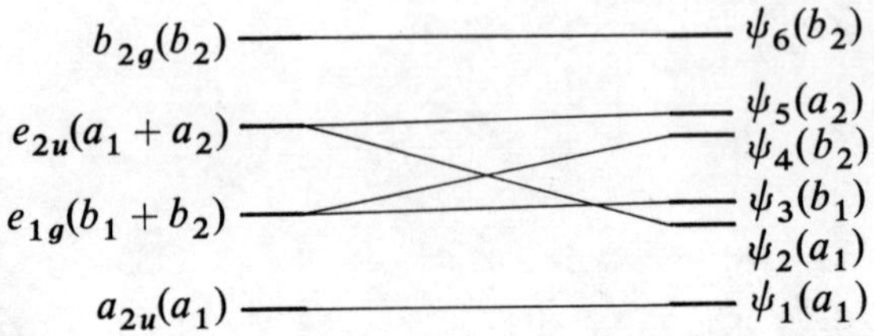


(c)

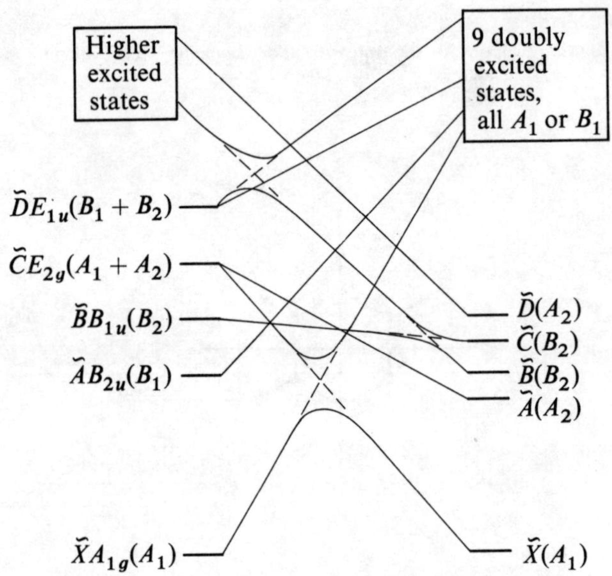


(d)

**Fig. 4.23**

(a) Electronic configurations of the states of Dewar benzene.

(b) Symmetrized MO's of Dewar benzene. For antibonding orbitals, the positive lobe (in the positive $z$ direction for $\pi^*$ and along the bond axis for $\sigma^*$) is taken to be near the C atom of lower number.

(c) Orbital correlation diagram for the transformation, benzene → Dewar benzene.

(d) State correlation diagram for the transformation, benzene → Dewar benzene.

uch a species, symbolized as

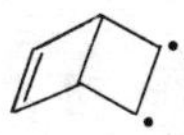

ay react in several ways.

(a) Radiationless decay to the ground state, giving Dewar benzene itself as product.

(b) Reaction with an olefin to give **17**. This transformation is at present only known in hexafluorobenzene.

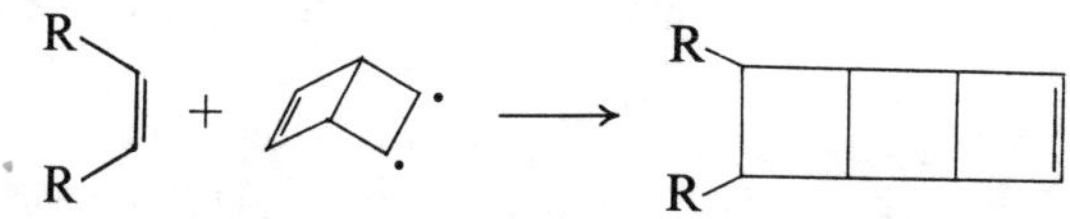

**17**

(c) Reaction to give prismane. Benzene itself is not known to be transformed into prismane although the reaction does occur in certain poly *t*-butyl-substituted derivatives.

The $^3B_{1u}$ state of benzene may distort so as to correlate[50] with the lowest plet state of a second species,

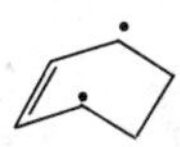

ich may also react with pyrrole[42] to give **18**:

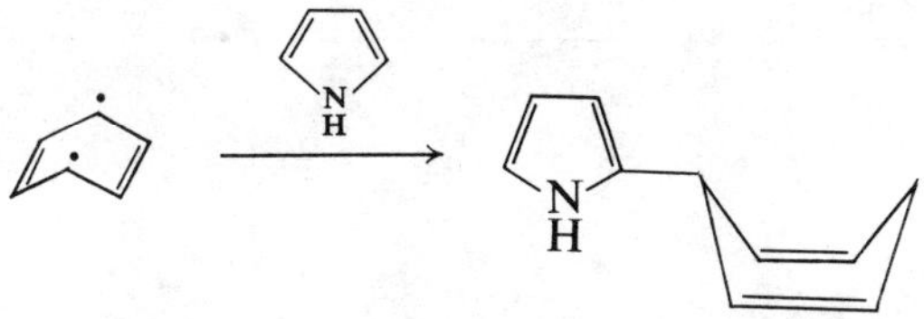

**18**

ese photoinitiated transformations of benzene are summarized in Fig. 4.24. Dewar benzene itself has been made by an indirect route,[52] and undergoes ence isomerization to benzene by a forbidden (Fig. 4.23, part d) and, therefore,

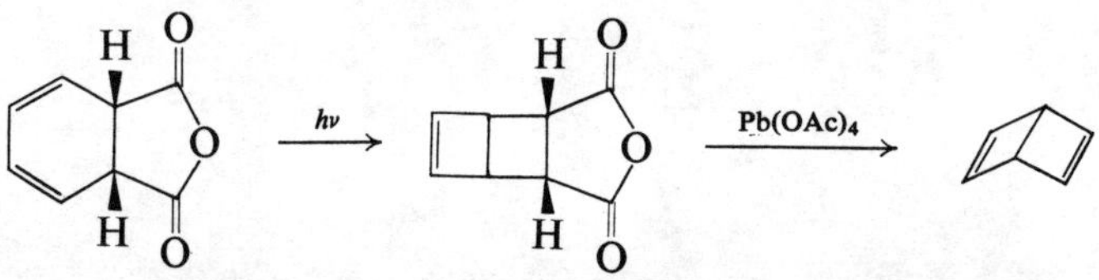

**103**

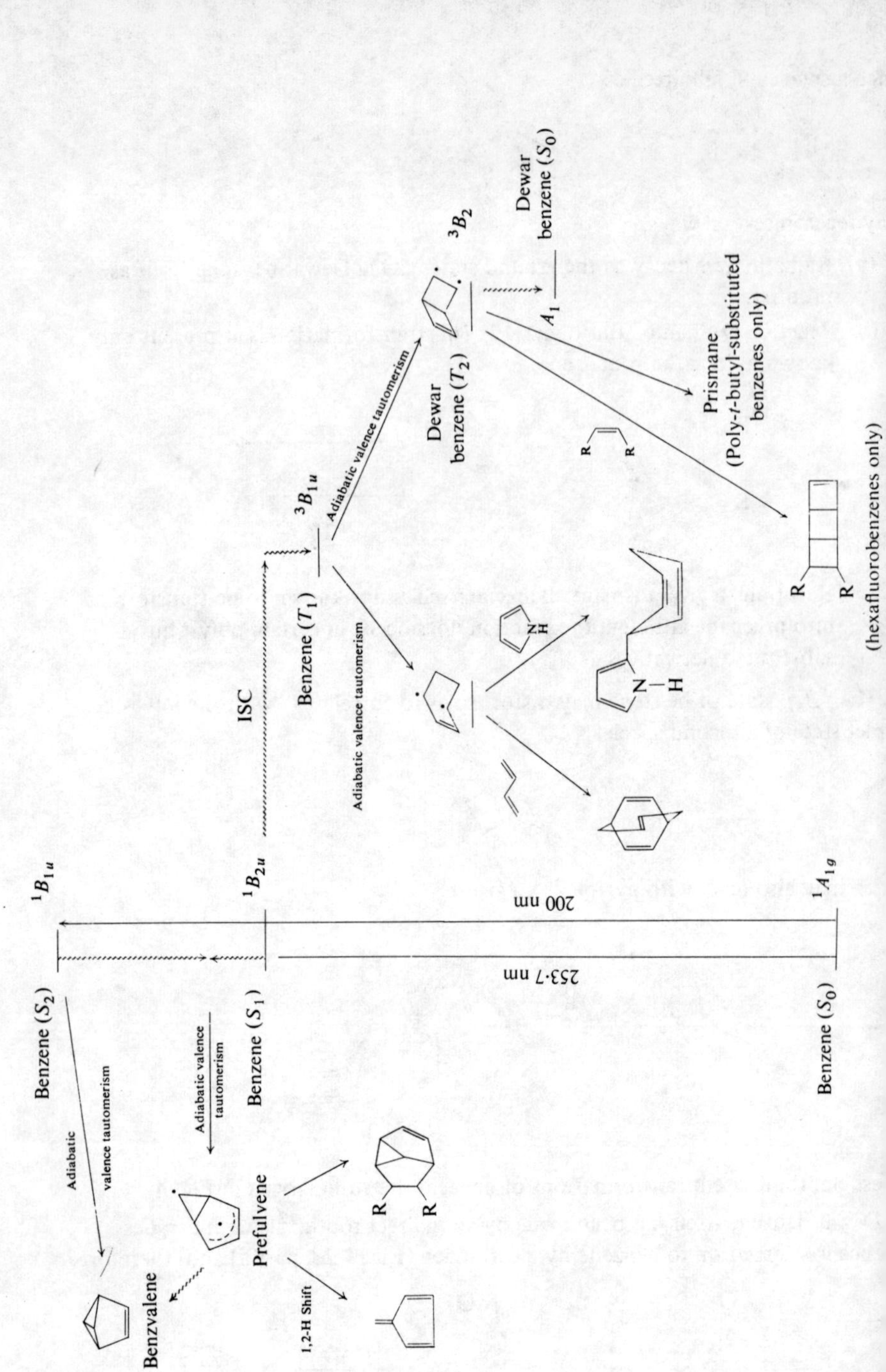
Benzvalene
Adiabatic
valence tautomerism
Benzene ($S_2$)
$^1B_{1u}$
Adiabatic valence
tautomerism
Benzene ($S_1$)
$^1B_{2u}$
Prefulvene
1,2-H Shift
R
R
ISC
Benzene ($T_1$)
$^3B_{1u}$
Adiabatic valence tautomerism
Adiabatic valence tautomerism
Dewar
benzene ($T_2$)
$^3B_2$
$A_1$
Dewar
benzene ($S_0$)
Prismane
(Poly-t-butyl-substituted
benzenes only)
(hexafluorobenzenes only)
N—H
H
N
200 nm
253·7 nm
Benzene ($S_0$)
$^1A_{1g}$

slow process ($t_{1/2}$ = 2 days at room temperature in pyridine) thus accounting for its stability. Van Tamelen[53] has pointed out that the transformation, Dewar benzene → benzene, either breaks the Woodward–Hoffmann rules, which is the essence of the above rationale of its stability, or leads to the highly strained *cis,cis,trans*-cyclohexatriene.

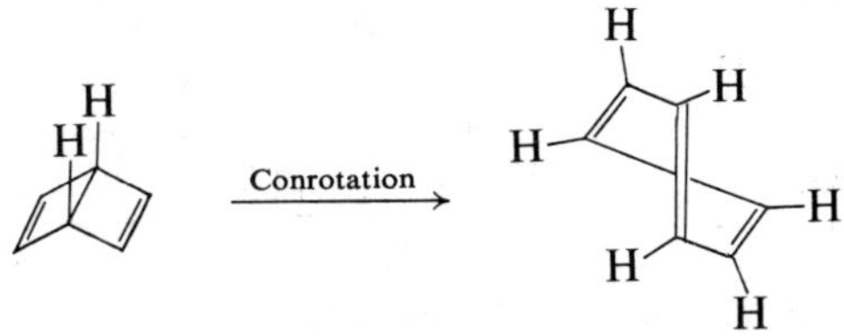


In passing, it is interesting to note that *cis,cis,trans*-cyclohexatriene is often referred to as Möbius benzene,[54] as the signs of the *p*-lobes on each face of the ring have one discontinuity, and thereby assume the topology of a Möbius strip.

### *4.2.3 Dienones and related compounds*[55]

Irradiation of 4,4-diphenylcyclohexadienone results in the formation of 2,3-diphenylphenol and 6,6-diphenylhexa-3,5-diencarboxylic acid, via the intermediacy of 6,6-diphenylbicyclo[3.1.0]hex-3-en-2-one.

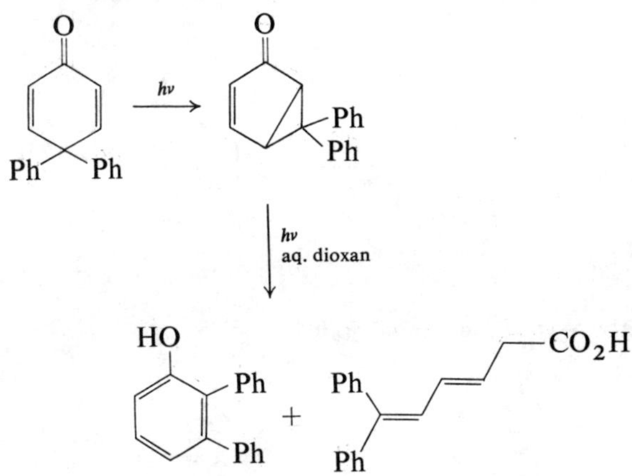


Zimmerman,[56] who has investigated this reaction, sees it as proceeding through the following stages.

(a) $\pi^* \leftarrow n$ Excitation giving the $(n, \pi^*)$ state.

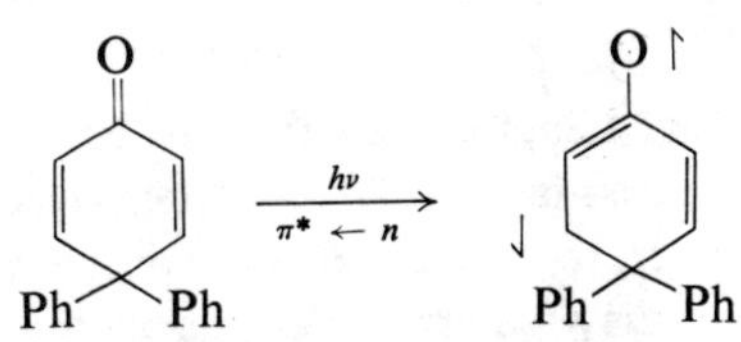


Only one resonance contributor is shown.

(b) Intersystem crossing. This is assumed since acetophenone sensitizes the reaction:

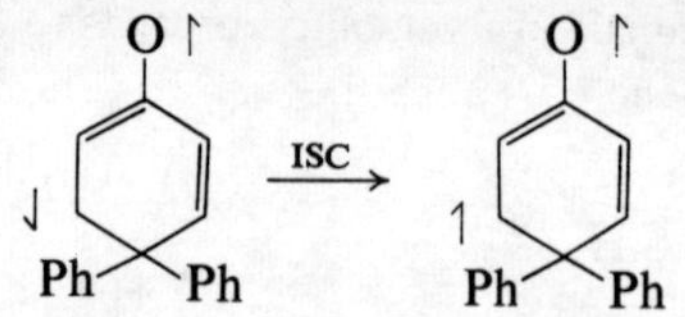


(c) Rearrangement of the excited state.

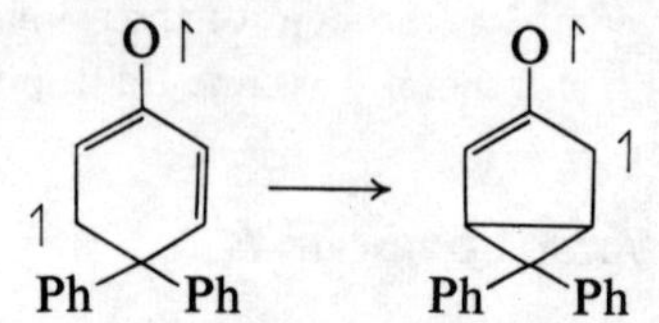


(d) $\pi^* \rightarrow n$ Electron demotion to give a ground state species.

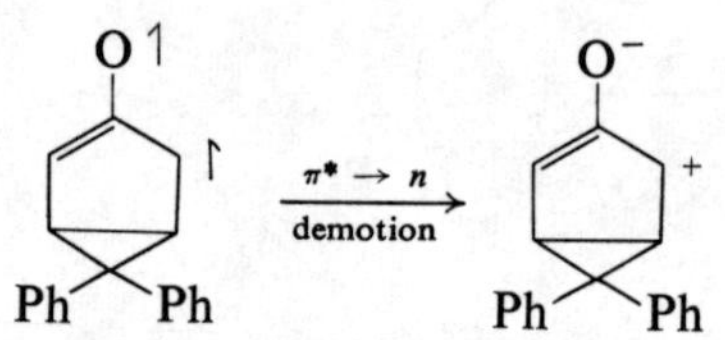


(e) Rearrangement of the ground state.

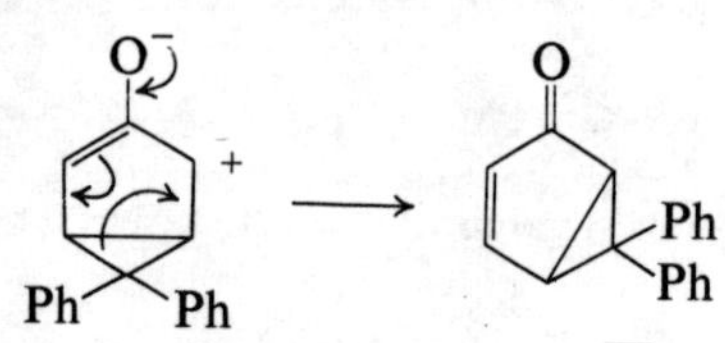


The MO's are shown in Fig. 4.25, from which it can be seen that $\pi^* \leftarrow n$ promotion leads to an increase in the 3,5-bond order, in accordance with stage (c). It is important, however, to realize that such considerations as these must be treated with reserve, since they require localization of the electrons into $n$, $\pi$, and $\sigma$ orbitals, a process which is not necessarily valid.

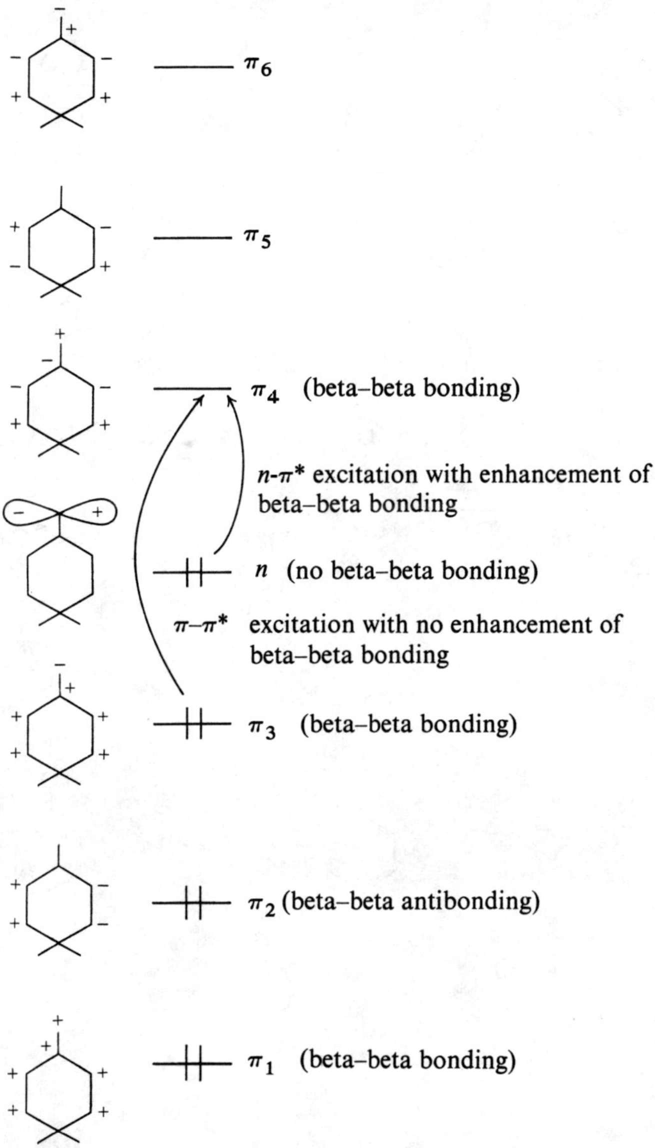


**Fig. 4.25** Molecular orbital representation of $\pi^* \leftarrow n$ and $\pi^* \leftarrow \pi$ excitation processes for cyclohexadienones. +, indicates wavefunctions positive. −, indicates wavefunctions negative.

Application of the same approach to the intermediate bicycle leads to the products as follows.

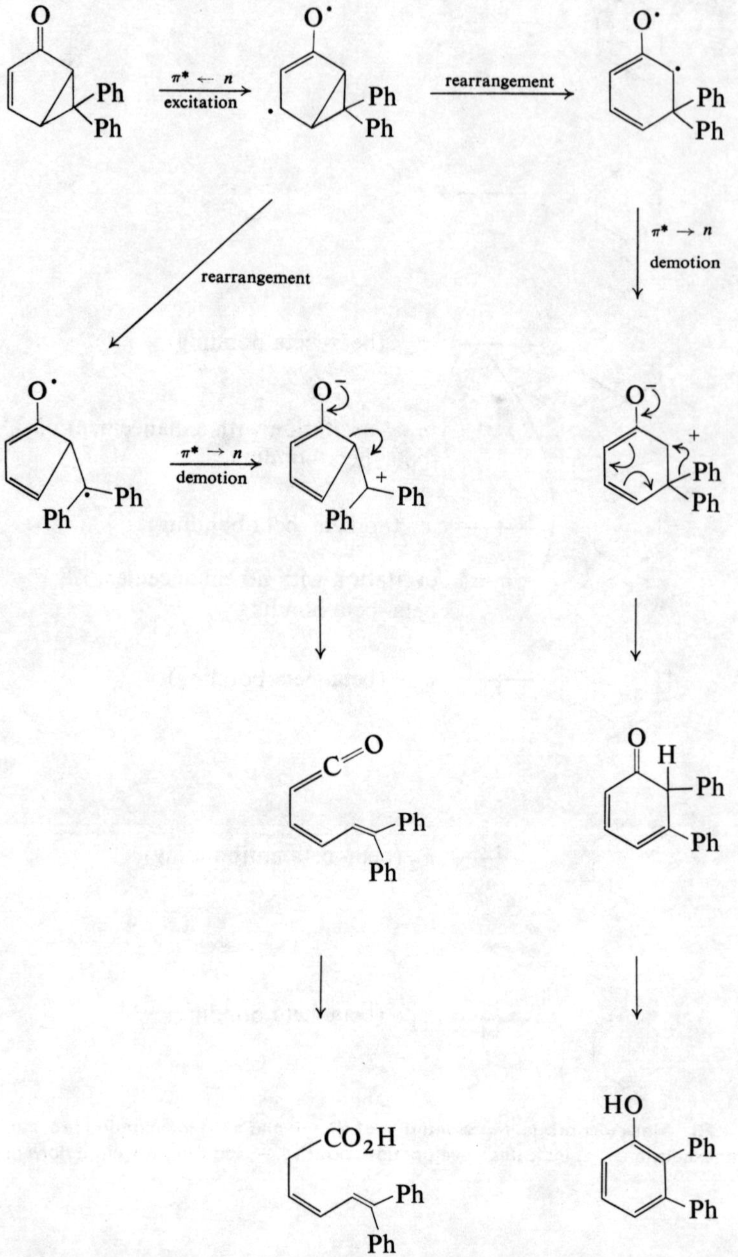

The photochemistry of santonin has been investigated,[55, 56] the major products of irradiation being illustrated below.

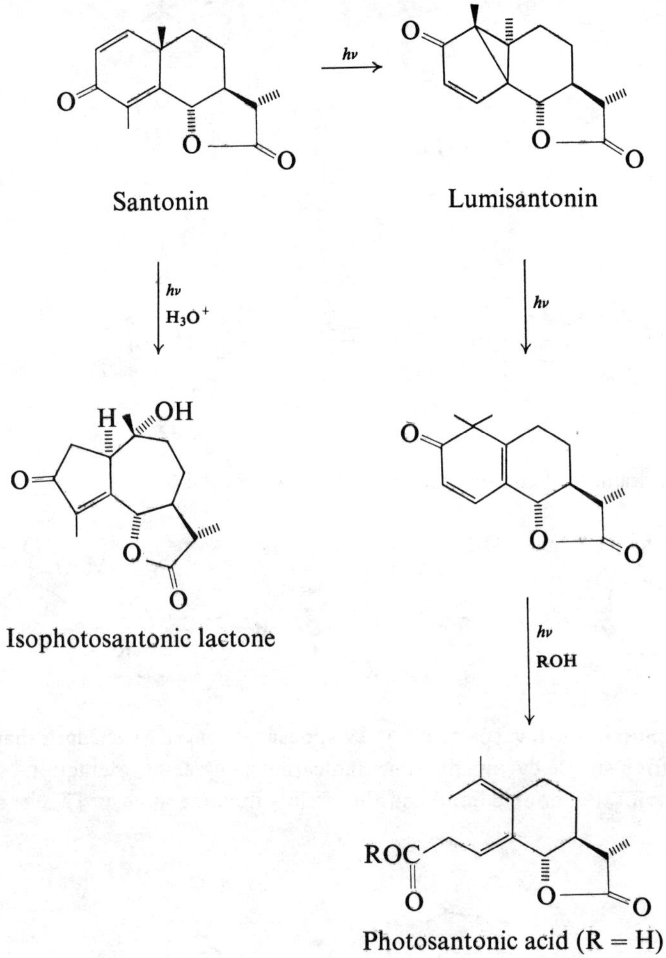


The photorearrangements of tropones and derivatives have been the subject of a good deal of attention and have been reviewed.[30] Simple valence isomerization can occur,

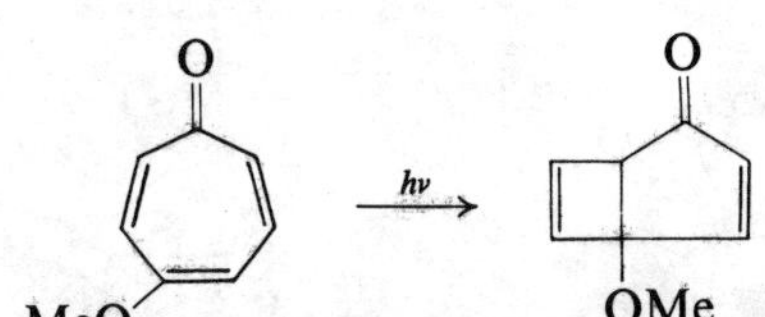

or the reaction can be more complex,[57]

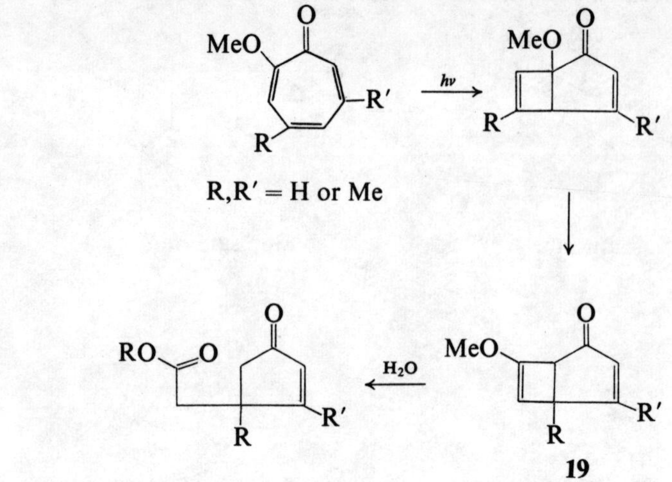


The mechanism of the first step has been represented by,

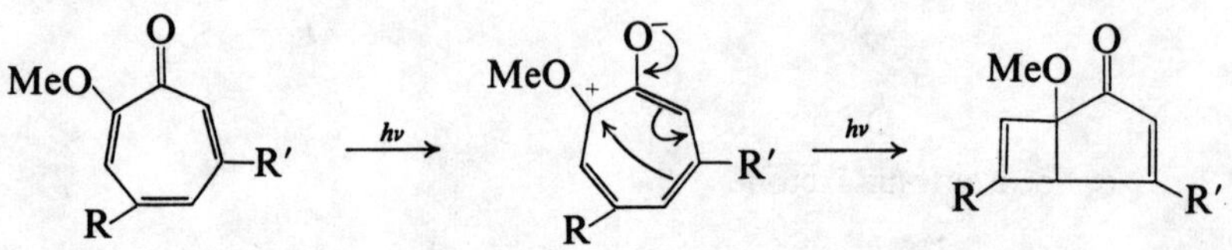


The maximum in the u.v. spectrum of **19** appears at longer wavelength than expected for a simple cyclopentenone, indicating a degree of interaction between the non-conjugated double bond and the cyclopentenone moiety. This leads to,

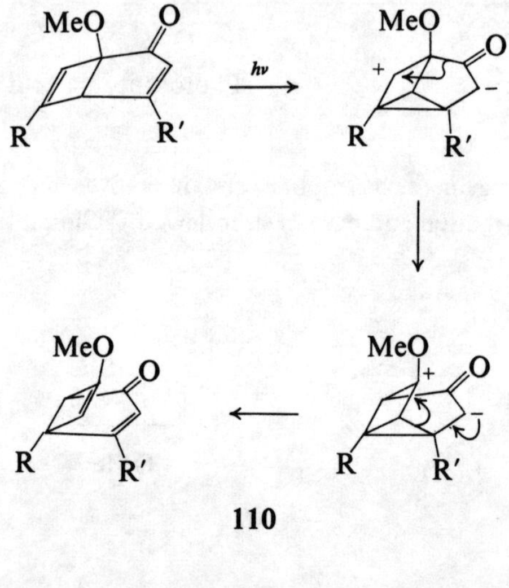


**110**

### 4.2.4 Photo-Fries[58] and photo-Claisen rearrangements

The photochemical analogue of the Fries rearrangement is known and is believed[59] to involve an upper singlet state, and a 1,3-sigmatropic shift.[1] More recently, evidence has been presented suggesting a radical cage process:[60]

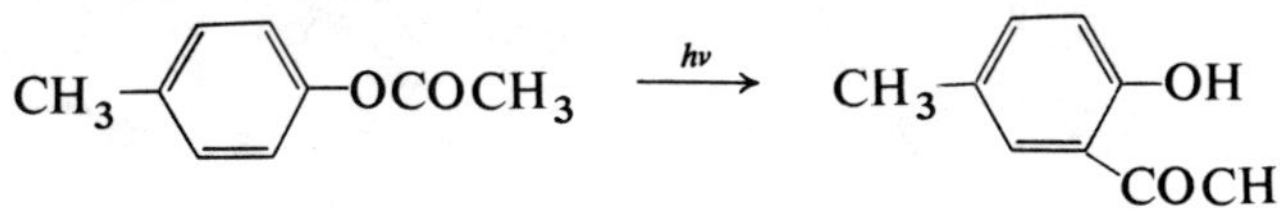


This reaction has found utility in the synthesis of griseofulvin:

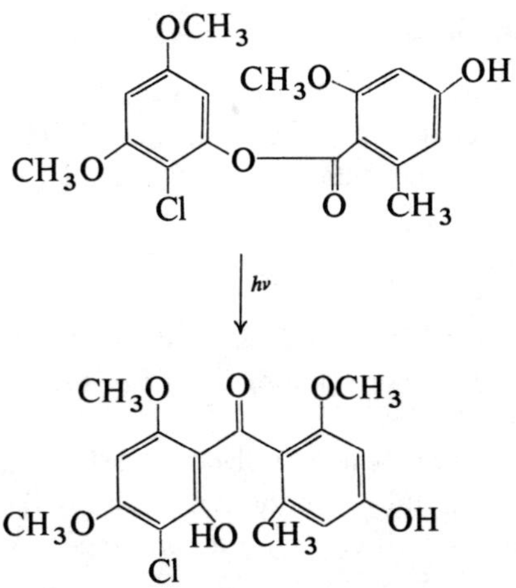


The photo-Claisen reaction is also known:[61]

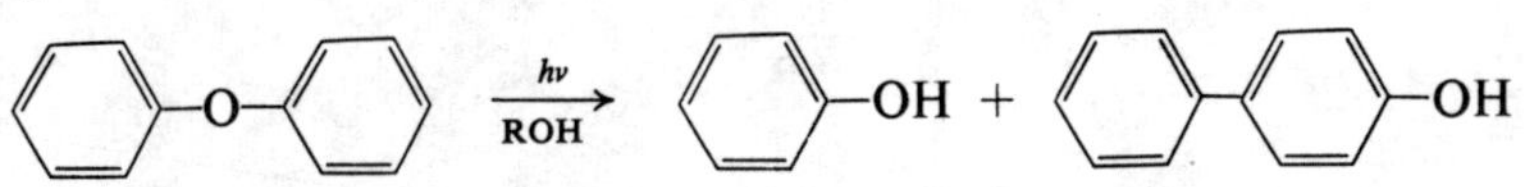


### 4.2.5 Stilbene cyclization

It is often convenient to consider excited stilbenes as composed of two major resonance contributors,

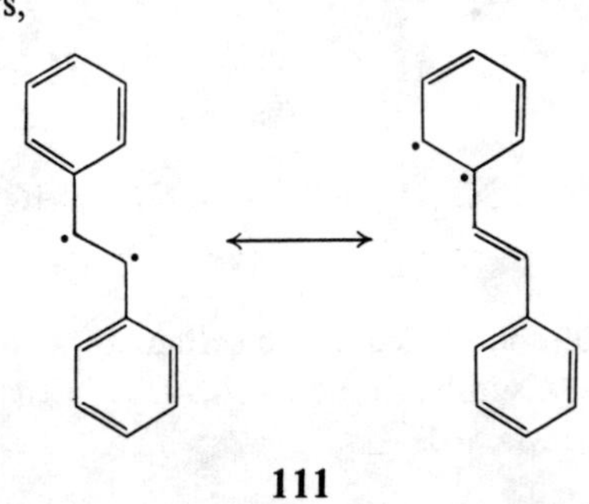

although in many cases such localization of the excitation will not be valid. Irradiation of *trans*-stilbene in a variety of solvents leads to phenanthrenes[45] via an excited singlet state and the intermediacy[46] of a *trans*-dihydro compound:

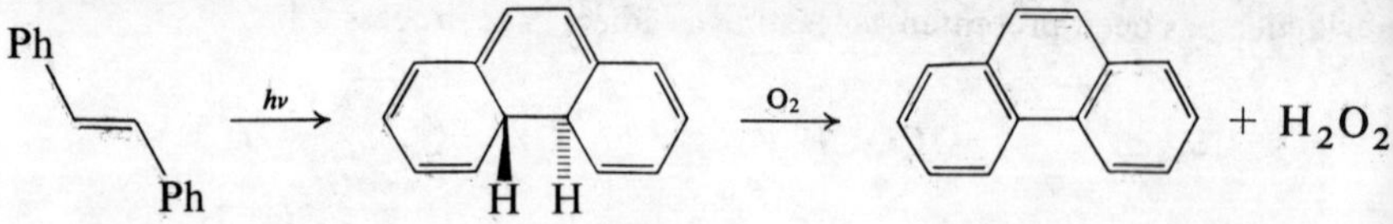

Azobenzene will also undergo this type of oxidative cyclization in acidic media.[6

## 4.3 Addition reactions

Photochemical addition reactions are known in all three phases; they can be inter- or intramolecular and may often proceed with ring formation. The latter, if concerted, are subject to orbital symmetry control.[1]

### *4.3.1 Simple addition to olefins*[65]

Under this heading, only those reactions are discussed which do not proceed with ring formation. The photoaddition of halogens to olefins is mentioned in section 3.2.5.

Because of its possible biochemical significance,[66] the photoaddition of water to pyrimidines has been studied, but hitherto, this reaction has had little importance as far as synthetic chemistry is concerned.

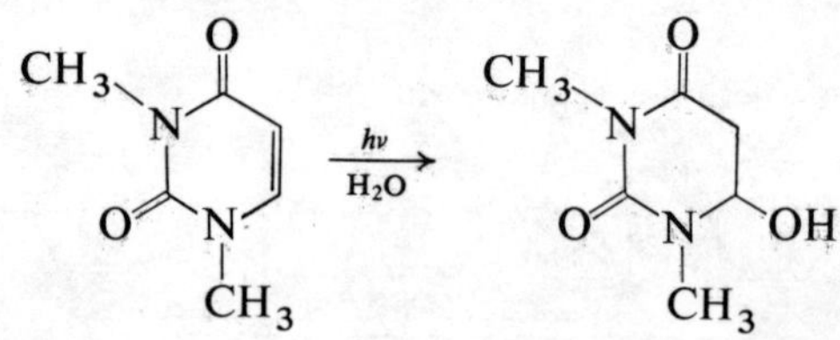

The photoaddition of alcohols to olefins has been found to proceed in two ways, leading either to a higher alcohol derived by homolysis of the α-CH bond of the substrate alcohol,[67] or to an ether by a polar pathway:[68]

$$CH_3(CH_2)_5CH{=}CH_2 \xrightarrow[\succ\!\!-OH]{h\nu} CH_3(CH_2)_7\overset{\displaystyle OH}{\overset{|}{C}}(CH_3)_2$$

$$CH_3CH{=}CHCO_2H \xrightarrow[CH_3OH]{h\nu} CH_3\underset{\displaystyle OCH_3}{\underset{|}{C}}HCH_2CO_2H$$

Photosensitized addition of alcohols to certain cyclic olefins has been developed and is of both synthetic and mechanistic interest.[69] *A priori* three limiting cases may be envisaged.

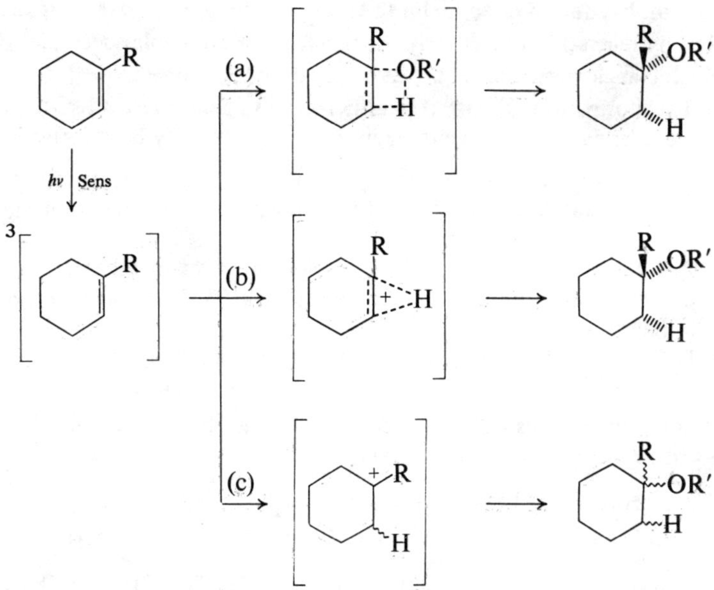


(a) A concerted 4-centre *cis*-addition.
(b) A *trans*-addition possibly via a protonium species.
(c) A stepwise addition via a cationic species.

Irradiation of 2,10-dimethyl-*trans*-2-octalin in the presence of $D_2O$ and xylene as sensitizer leads to the exocyclic isomer, **20**, and to the alcohols, **21**, and **22**, thereby providing strong support for mechanism (c).

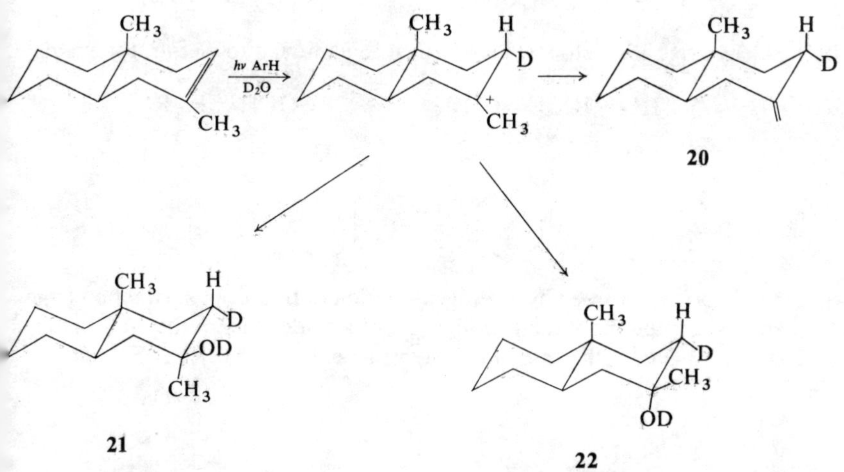


The reaction is limited to six- and seven-membered rings in that cyclopentenes yield products of free radical reactions while cycloöctenes either polymerize or undergo *cis-trans* isomerization. This behaviour is attributed to differences in steric strain. Excitation of an olefin to either a *cis* or *trans* triplet is rapidly followed by relaxation to the thermodynamically more stable species in which the dihedral angle between the $sp^2$ hybridized carbon atoms is 90° (cf. section 4.2.1). Transformation ($T_1 \rightarrow S_0$) to either the *cis* or *trans* ground state can then take place. Clearly three different reaction modes can easily be envisaged.

(a) As a consequence of the inability of small rings (e.g., cyclopentene) to accommodate orthogonal $p_z$ orbitals, the product of excitation cannot undergo vibrational relaxation.

(b) As a consequence of the inability of six- and seven-membered rings to accommodate a *trans* double bond, the pathway $T_1 \rightarrow S_0$ (*trans* geometry is inadmissible.

(c) Both *cis*- and *trans*-cycloöctenes (and higher homologues) are known, thu the transition $T_1 \rightarrow S_0$(*trans*) is geometrically possible.

Ammonia and amines will photoadd to olefins, leading in the case of butylamine and 1-octene to 4-aminododecane:[70]

$$CH_3CH_2CH_2CH_2NH_2 + C_6H_{13}CH{=}CH_2 \xrightarrow{h\nu} CH_3CH_2CH_2\overset{NH_2}{\overset{|}{C}}HCH_2CH_2C$$

It is believed that the reaction involves generation of the radical $R\dot{C}HNH_2$ which then adds to the olefin.

The work of Elad shows that a smooth photosensitized addition of formamide to terminal olefins can be achieved[71] via the intermediacy of the carbamoyl radical $\dot{C}OHN_2$:

$$RCH{=}CH_2 + HCONH_2 \xrightarrow[Me_2CO]{h\nu} RCH_2CH_2\overset{O}{\overset{\|}{C}}NH_2$$

Both aldehydes[72] and ketones[73] are known to photoadd to olefins, for example,

$$RCHO + R'CH{=}CH_2 \xrightarrow{h\nu} RCOCH_2CH_2R'$$

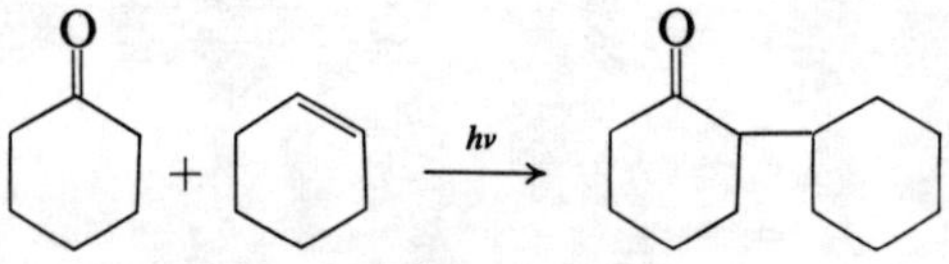

The reactions may proceed by hydrogen ejection or hydrogen abstraction from the carbonyl compound leading, in the case of an aldehyde, to the nucleophilic acyl radical 23 and, in the case of cyclohexanone, to the species 24.

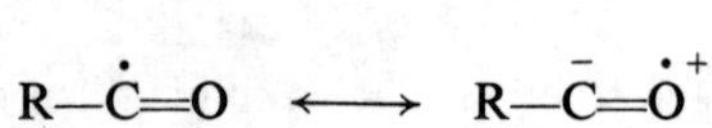

**23**

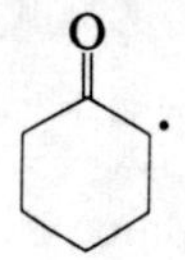

**24**

The formation of cyclohexanol as a by-product in the latter reaction is an indication that the second pathway might be operating.

### 4.3.2 *Cycloaddition reactions*

The concept of the conservation of orbital symmetry (p. 87) already applied to electrocyclic reactions and sigmatropic rearrangements, can be extended to include *concerted* cycloaddition reactions. First, consider the simple case of the dimerization of two ethylene molecules to give cyclobutane (the $2\pi + 2\pi$ cycloaddition). It is assumed that immediately preceding reaction, the two ethylene molecules orient themselves in parallel planes directly above each other (Fig. 4.26).

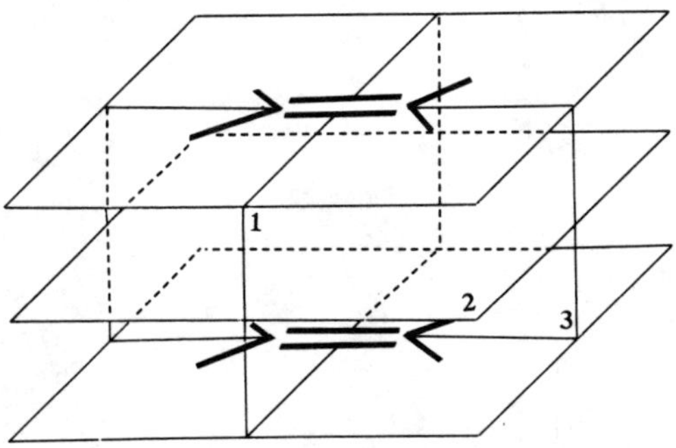


**Fig. 4.26**

Such a disposition shows three planes of symmetry (Fig. 4.26) according to two of which it is convenient to classify the orbitals directly involved in the reaction. These are the $\pi$ and $\pi^*$ orbitals of the reactant ethylene molecules and the $\sigma$ and $\sigma^*$ orbitals of the product cyclobutane molecule. The symmetries of the orbitals are now classified according to planes 1 and 2 only (Fig. 4.26), since plane 3 is a symmetry element according to which all the orbitals are symmetric and is of no use in deciding whether or not the reaction is symmetry allowed. The MO's for cyclobutane and for the combination of the two ethylene molecules are shown in Fig. 4.27 and Fig. 4.28, respectively.

Other combinations are invalid since they break the quantum mechanical principle that acceptable MO's must either be symmetric or antisymmetric with respect to any given symmetry element of the molecule. In reactions, the appropriate symmetry elements are those present in the transition state.

The correlation diagram for such a reaction is shown in Fig. 4.29 from which it can be seen that the symmetries of the orbitals of two ground state ethylene molecules correlate with the symmetries of the orbitals of a doubly excited form of cyclobutane. Thus, there is a large symmetry imposed barrier to the transformation.

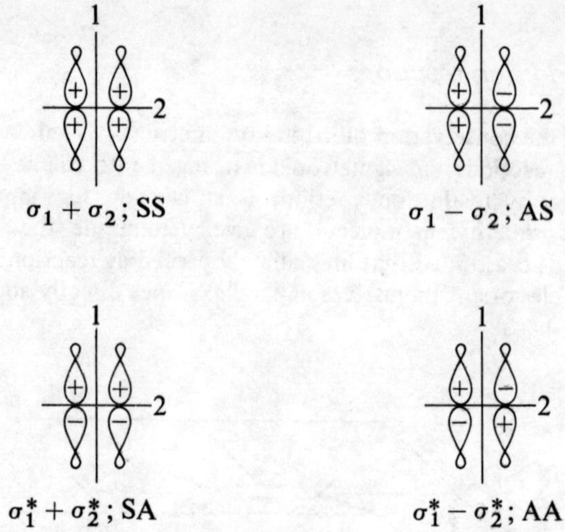


**Fig. 4.27**

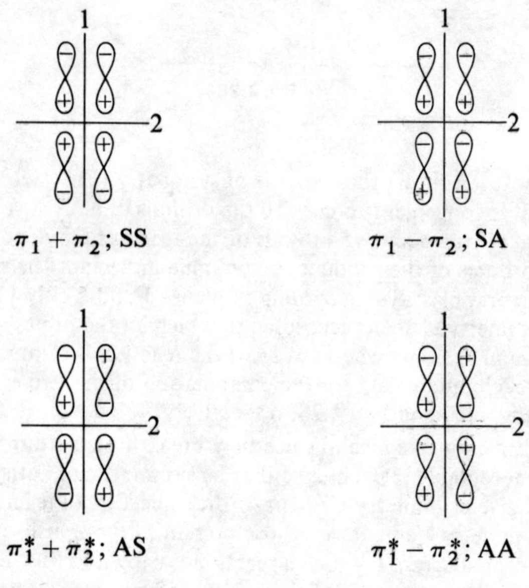


**Fig. 4.28**

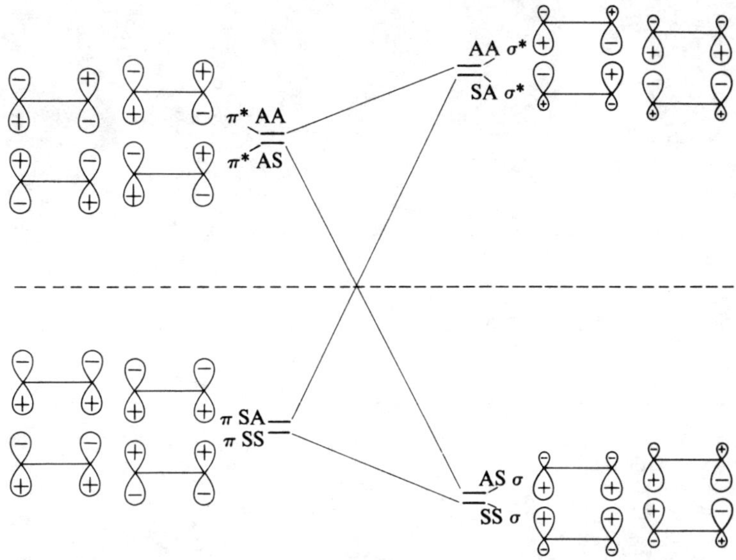


Fig. 4.29

A similar diagram drawn for the corresponding photochemical reaction shows that there is a correlation between the symmetries of the reactant orbitals and the symmetries of the orbitals of excited cyclobutane. This transformation is, therefore, symmetry allowed.

Precisely similar arguments can be used to develop a correlation diagram for the reaction between ethylene and butadiene (the $2\pi + 4\pi$ cycloaddition). The most reasonable geometry of approach possesses a plane of symmetry as shown in Fig. 4.30.

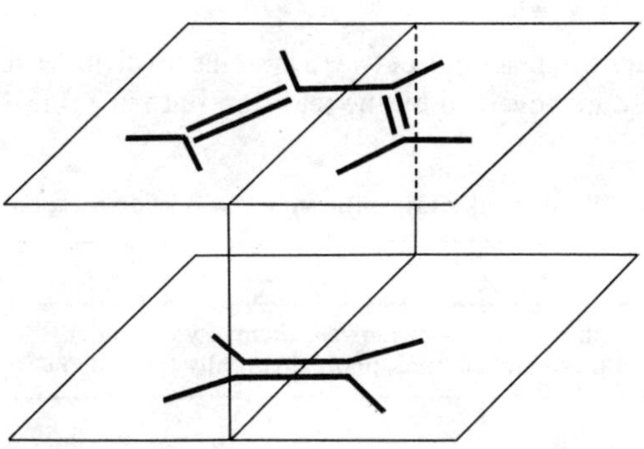

Fig. 4.30

The symmetries of the reactant and product orbitals are now classified with respect to this plane and construction of the correlation diagram (Fig. 4.31) shows that a ground state correlation exists between them. The reaction is, therefore, a symmetry allowed process. No correlation exists between the symmetries of the reactant and product orbitals in the first excited state, rendering the process photochemically forbidden.

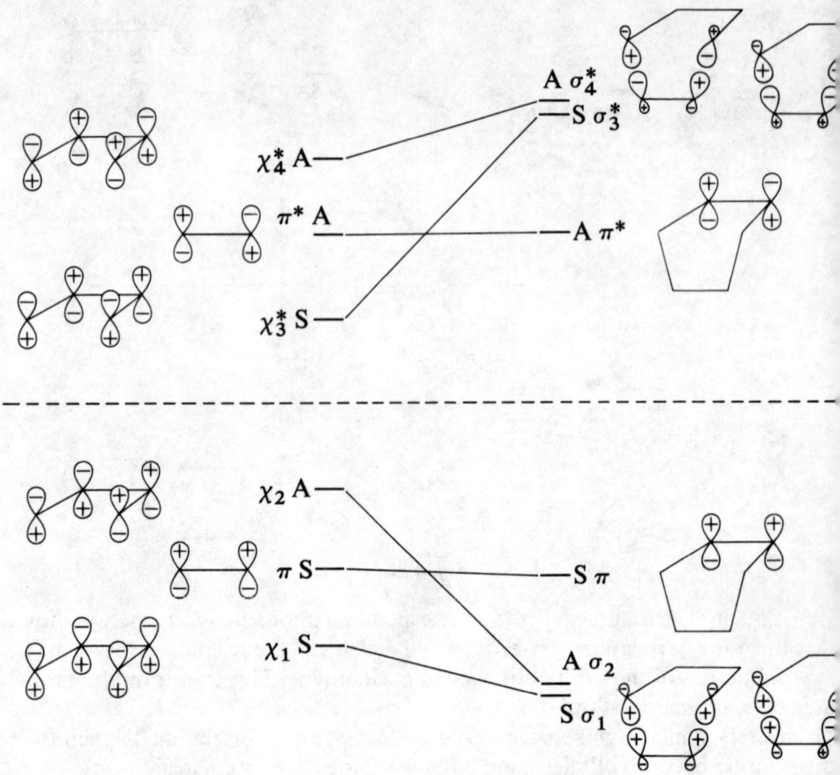


Fig. 4.31

In general, cycloaddition reactions between systems involving $p$ and $q$ $\pi$-electrons respectively are governed by the selection rules in Table 4.4.

TABLE 4.4: Selection rules for concerted cycloaddition reactions

| $p + q$ | |
|---|---|
| $4n$ | photochemically allowed, thermally forbidden |
| $4n + 2$ | thermally allowed, photochemically forbidden |

### *4.3.3 Photodimerization of simple olefins*

Direct irradiation of monoölefins gives rise to high energy singlets which undergo intersystem crossing only inefficiently. The use of sensitizers, benzene or ketones renders their triplets quite accessible and despite rapid degradation to the ground state, *cis-trans* isomerization is an important process (cf. p. 82).

If sensitized irradiations are carried out in high concentrations, some olefins ndergo dimerization,[74, 75] for example,

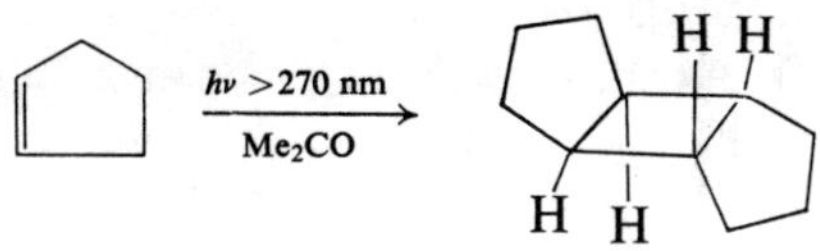


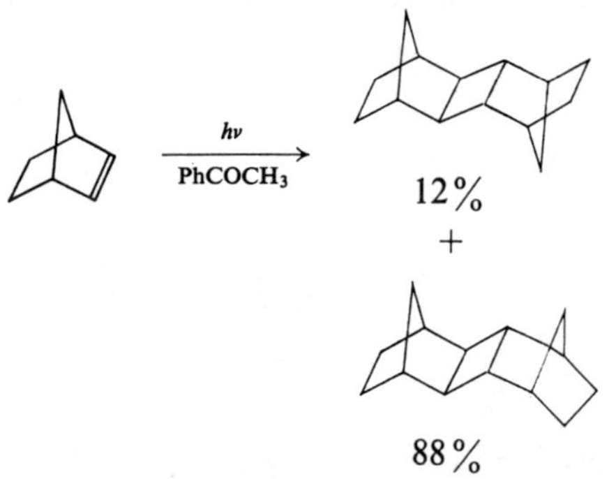


The photosensitized ($Ph_2CO$) intramolecular cycloaddition of norbornadiene ading to quadricyclane is a well known reaction (cf. p. 96):

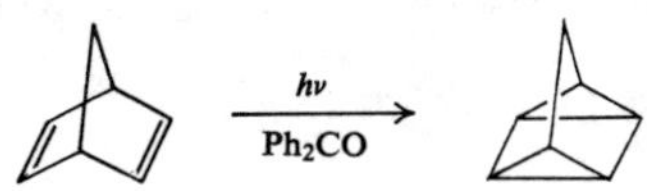


eactions of this type involving triplets are not concerted and consequently e not subject to the Symmetry Selection Rules.

### 3.4 *Photodimerization of conjugated dienes*

s with simple monoölefins, the singlet state of conjugated dienes is of high ergy, intersystem crossing inefficient, and consequently, diene triplets are ually generated by photosensitization. Irradiation of concentrated solutions dienes in the presence of a photosensitizer is a process equivalent to generatg diene triplets in the presence of high concentrations of diene, and results in action of the diene triplets with the diene, leading to biradical intermediates, nich can then undergo ring closure.[76]

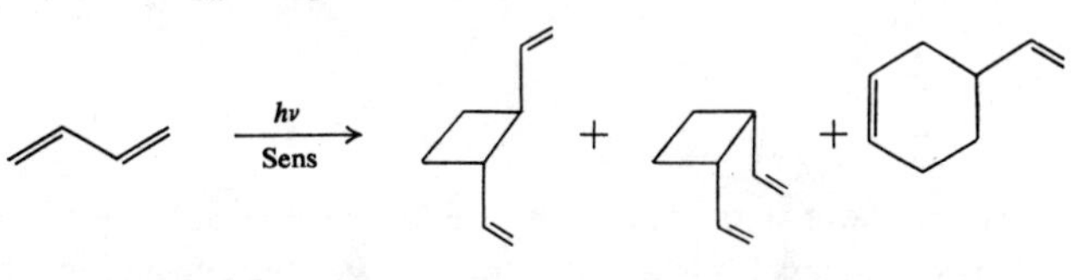


The product distribution has been shown by Hammond[77] to be a function of nsitizer triplet energy (cf. p. 84), low concentrations of cyclohexenes being rmed with sensitizers of $E_T > 251$ kJ mol$^{-1}$ and $E_T < 209$ kJ mol$^{-1}$. It is known that 1,3-cyclohexadiene has an $E_T$ of about 222 kJ mol$^{-1}$ and the results are interpreted as meaning that for sensitizers having a high $E_T$, excitation of the *s-trans* form is taking place, whereas below 251 kJ mol$^{-1}$ quenching by the *s-trans* form is inefficient and only the *s-cis* monomers in solution are excited. The conclusion drawn is that the two ground state conformers give rise to geometrically different triplets which react giving different products. The triplets resulting from the *s-trans* conformer lead to cyclobutanes and those from the *s-cis* lead to cyclohexenes.

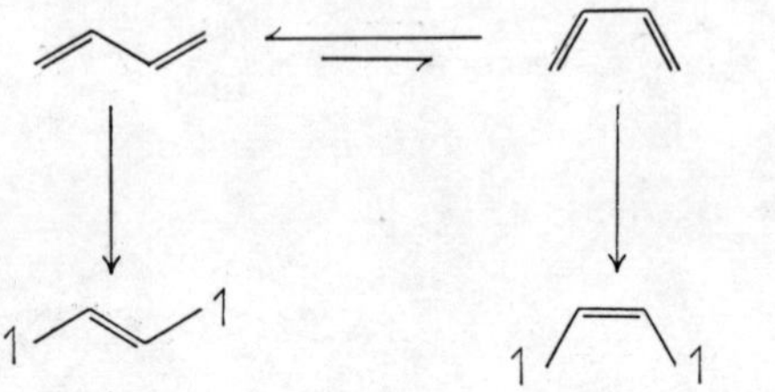

$E_T \sim 251$ kJ mol$^{-1}$ $\qquad$ $E_T \sim 222$ kJ mol$^{-1}$

It must be stressed that these two triplets once generated are non-interconvertible.

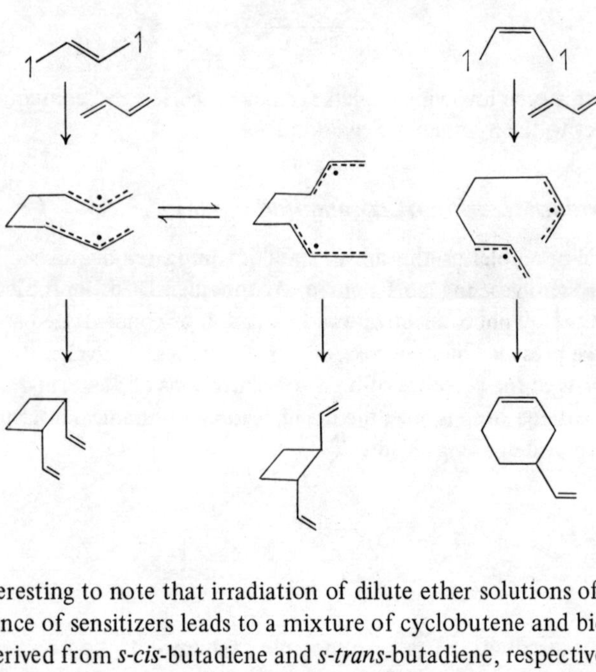

It is interesting to note that irradiation of dilute ether solutions of butadiene in the absence of sensitizers leads to a mixture of cyclobutene and bicyclobutane possibly derived from *s-cis*-butadiene and *s-trans*-butadiene, respectively. Evidence exists, however, suggesting that bicyclobutane formation may be a two-step process.

Cyclic dienes such as 1,3-cyclohexadienes also undergo photosensitized dimerization,[78] for example,

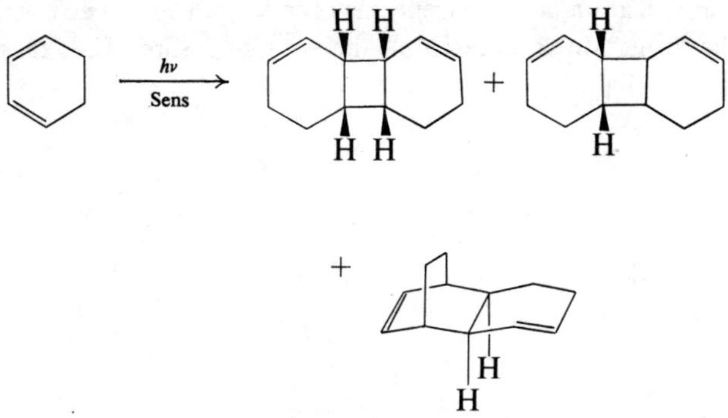


Cross additions of olefins have met with only limited success.[79] For example, no cross-adducts were obtained from irradiation of solutions of cyclopentadiene or 1,3-cyclohexadiene and 2-acetonaphthone, and olefins such as cyclohexene, norbornene, or 1,5-hexadiene. However, if a 15:1 mixture of cyclopentene and 1,3-cyclohexadiene is irradiated using 2-acetonaphthone as sensitizer, in addition to the photodimers of cyclohexadiene, three new compounds are formed, **25**, and the two stereoisomers of **26**.

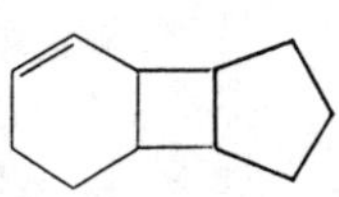

**25**

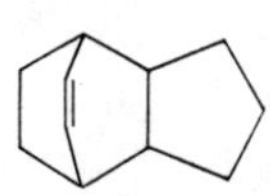

**26**

More interesting is the photochemistry of myrcene **27**. Direct irradiation leads to a complex mixture containing a cyclobutene **28** and β-pinene **29**,[80] whereas irradiation in the presence of a variety of sensitizers yields[79] a product formed by intramolecular cross-addition and formulated as **30**:

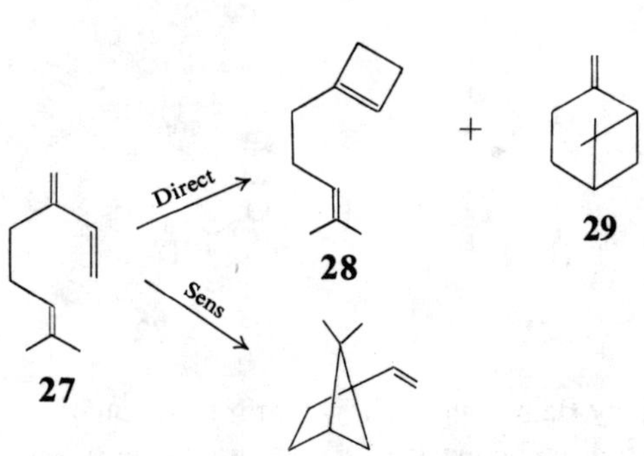

### 4.3.5 *Photodimerization of aromatic hydrocarbons and related compounds*

The photodimerization of anthracene has been known since 1867 when Fritzsche observed the formation of an insoluble dimer on exposure of benzene solutions of anthracene to sunlight:

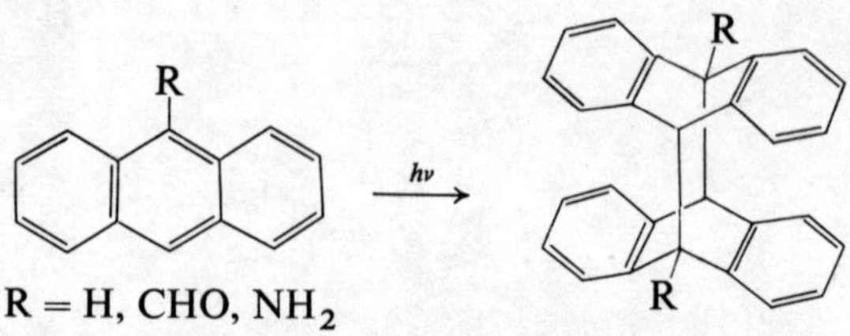


R = H, CHO, $NH_2$

9-Substituted anthracenes give rise to head to tail dimers.[82] The mechanism of the dimerization of the parent hydrocarbon has been the subject of much attention[83] and is believed to be:

$$A \xrightarrow{h\nu} {}^1A$$

$${}^1A + A \longrightarrow {}^1AA \text{ (Excimer; section 3.2.1)}$$

$${}^1AA \longrightarrow A_2$$

with competition from,

$${}^1A \longrightarrow A + h\nu_F$$

More recently 2-methoxynaphthalene has been found[84] to photodimerize, but dimerization of the hydrocarbon itself has so far not been achieved. In addition, a wide variety of polynuclear aromatic hydrocarbons have been found to photodimerize.[83]

Illumination of coumarin can lead to the cyclobutanes **31** and **32**:

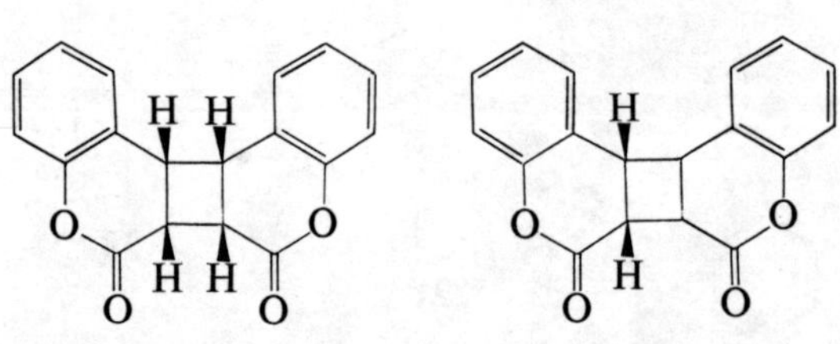


**31** **32**

Mechanistic studies by Hammond[85] show that interaction of excited state coumarin with ground state coumarin leads, in non-polar solvents, to self-quenching only and in polar solvents such as ethanol, to self-quenching, and the *cis* head to head dimer **31**, in very low yield. In the presence of triplet sensitizers, for example, benzophenone, the *trans* head to head dimer **32** is formed exclusively in both polar and non-polar solvents, even in those solutions in which most of the light is absorbed by the coumarin. It is believed that this is due to quenching of the coumarin singlets by the benzophenone, thus producing benzophenone singlets, which after intersystem crossing, transfer the energy back to the coumarin, thereby generating coumarin triplets (Fig. 4.32). This is

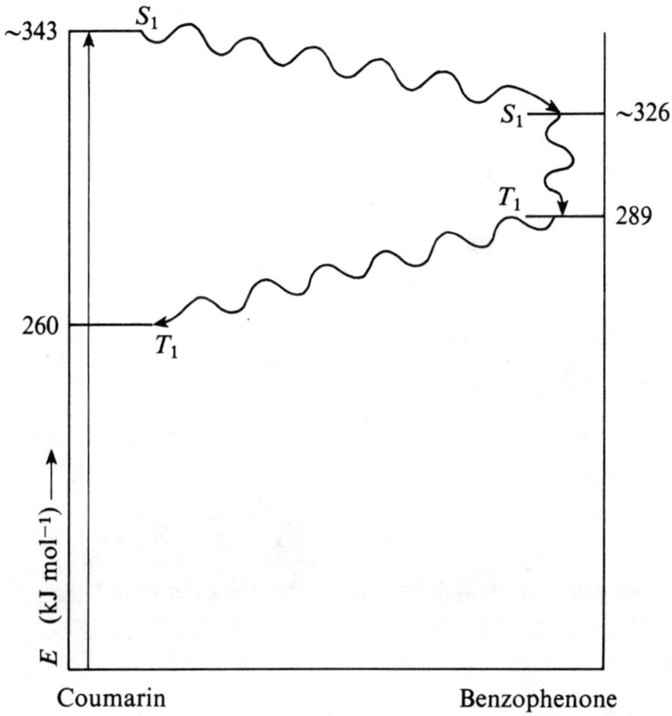


**Fig. 4.32** The low-lying excited states of coumarin and benzophenone. The fate of excitation absorbed originally by coumarin is traced by the arrows.

a nice example of a compound acting both as a singlet quencher and triplet sensitizer. The closely related 2-pyrone photoadds[86] internally,

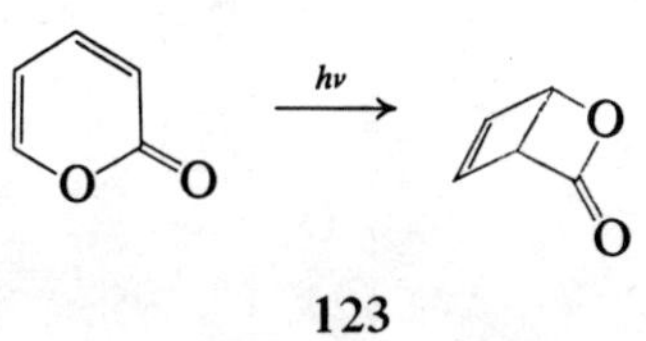

while the 4,5-diphenyl substituted compound yields 1,2,4,7-tetraphenylcycloöctatetraene, **33**, *p*-terphenyl, **34**, and diphenylacetylene, **35**. The following reaction sequence has been suggested[87] to account for this transformation:

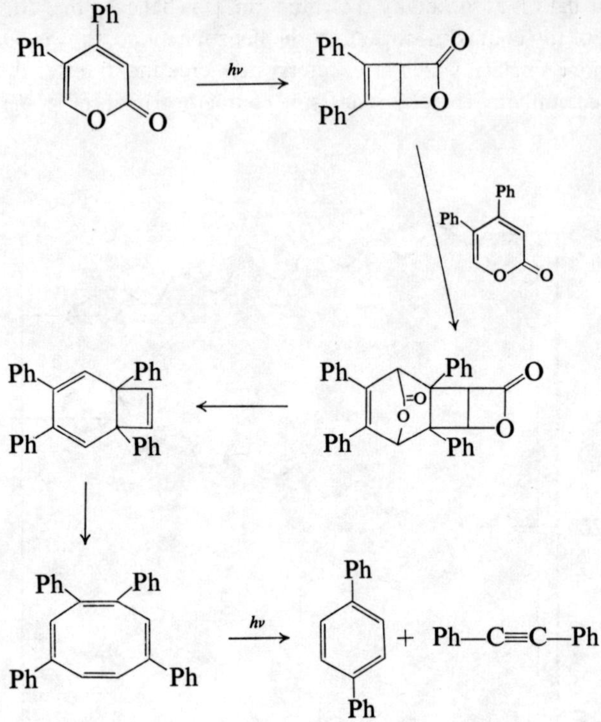


**33** **34** **35**

Both 2-aminopyridines and 2-pyridones, however, dimerize.[88]

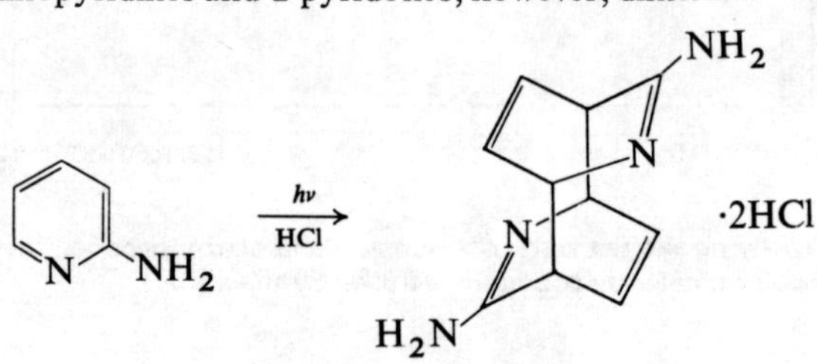


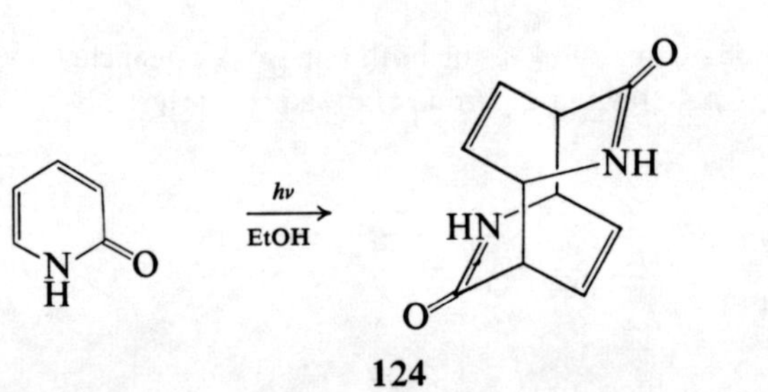


**124**

### *4.3.6 Dimerizations in the solid state*

The photodimerization of cinnamic acid has been known for many years, but only recently has a discussion of the factors affecting this reaction been forthcoming. Generally, one of the controlling factors in solid state organic reactions is the geometry of the crystal structure of the reactant species, as shown by,[89]

(a) chemically closely related species displaying significant differences in photochemical behaviour in the solid state,
(b) given compounds reacting differently in the solid and dispersed phases,
(c) polymorphic modifications of a given compound exhibiting significant differences in photochemical behaviour.

Reaction in the solid state occurs with a minimum of atomic or molecular movement, and consequently *trans*-cinnamic acid, which is dimorphic, may give rise to two different photodimers[90,91] because of two differing nearest neighbour orientations:

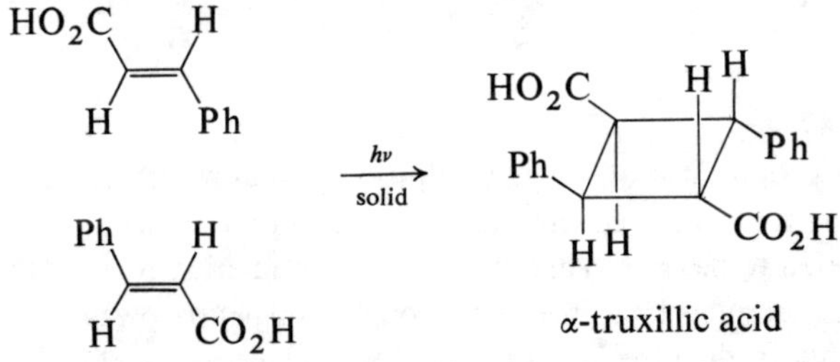


α-form

α-truxillic acid

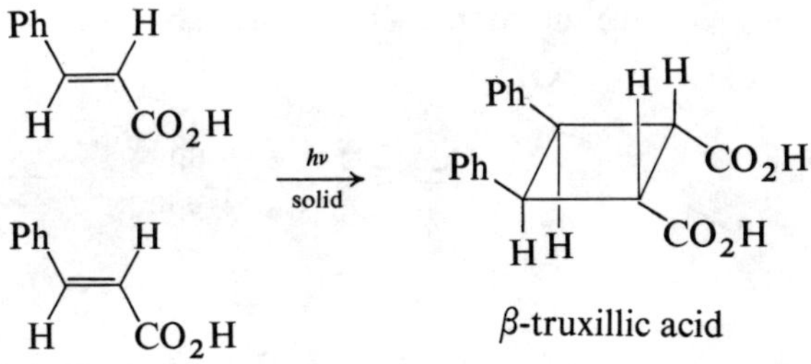


β-form

β-truxillic acid

Solid state irradiation of 2,3-dimethyl-1,4-benzoquinone leads[92] to a dimer, **36**, which on further irradiation forms a cage structure, **37**.

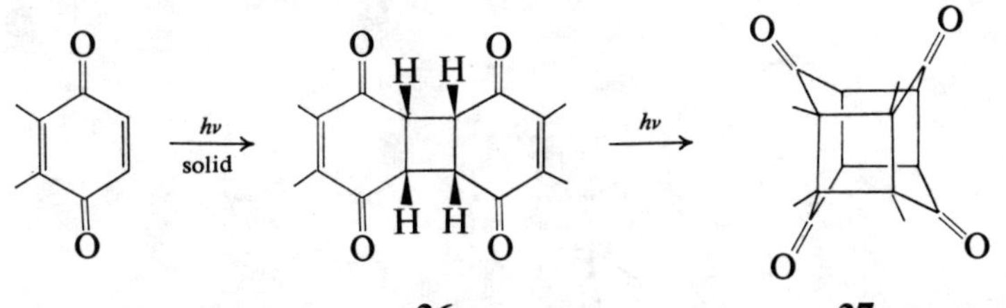


**36** **37**

### *4.3.7 Addition of maleic acid derivatives*

The photoaddition of derivatives of maleic acid to olefins is a synthetically useful and mechanistically well understood reaction. For example, irradiation[93] of the dimethyl ester in cyclohexene leads to allylic addition products, **38**, and, **39**, and also to a series of bicyclo[4.2.0]octane dicarboxylic esters, **40-43**:

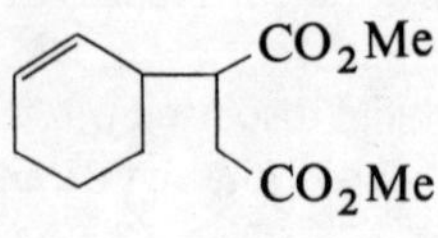


**38**

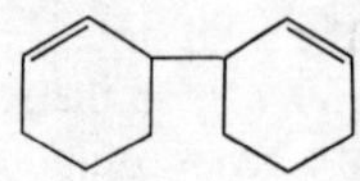

**39**

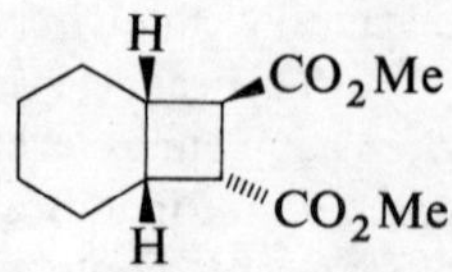


**40**

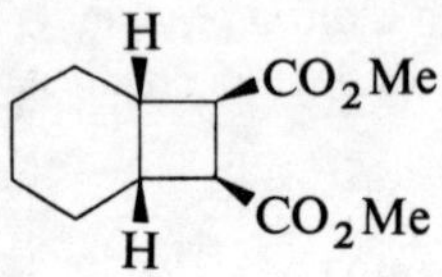


**41**

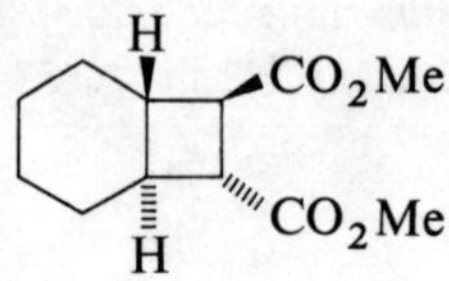


**42**

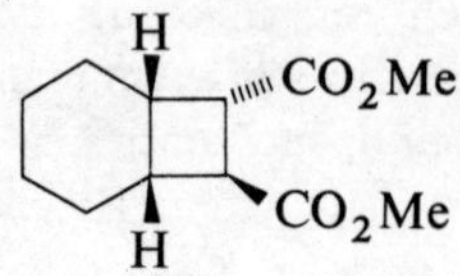


**43**

The relative proportions of the cycloaddition products vary according to the temperature and to whether the irradiation is direct or sensitized. The excited state is considered to be $(\pi, \pi^*)$ and in the direct irradiation it is believed that at least part of the products are generated through a singlet pathway.

Barltrop[94] has studied the reaction between cyclohexene and maleic anhydride and concludes that it proceeds by excitation of a charge-transfer band to give a biradical intermediate, which can collapse to either a *cis* or *trans* ring junction as well as the substitutive adducts, **44**, **45**:

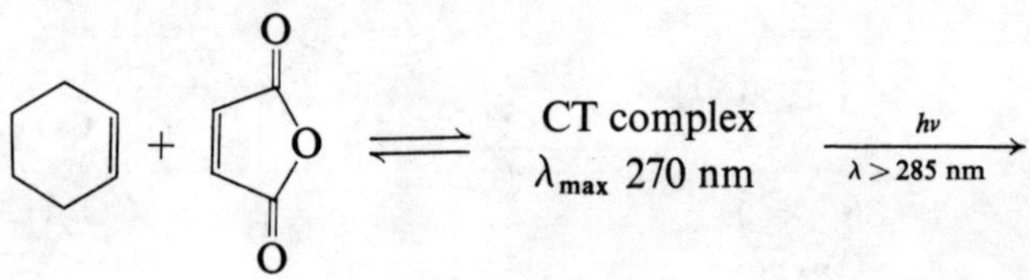


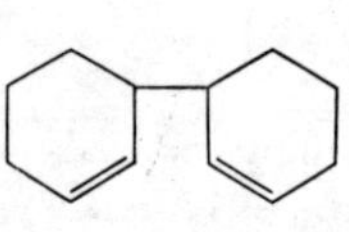

**44**

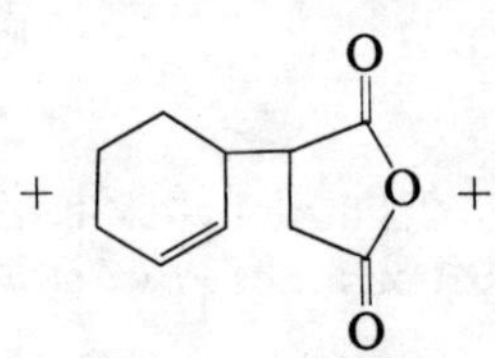


**45**

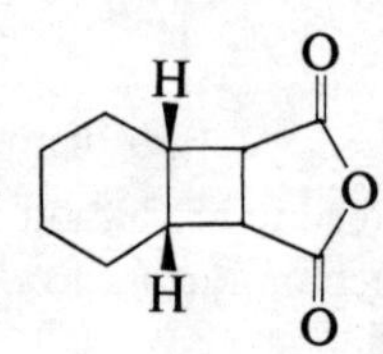


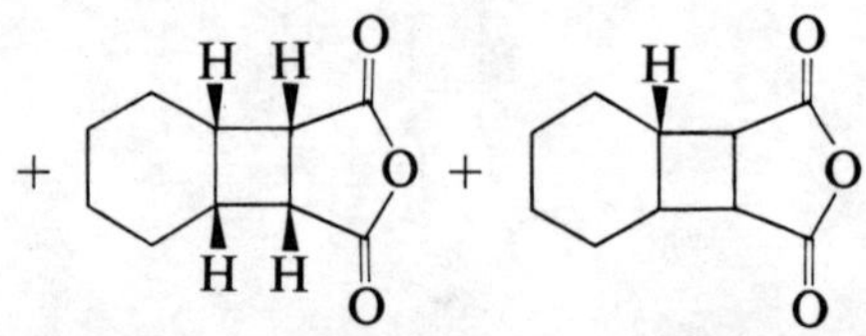

### 4.3.8 αβ-Unsaturated ketones and olefins

The photocyclization of αβ-unsaturated ketones to olefins has proved to be of general synthetic utility. A notable case is the synthesis of caryophyllene, **46**, reported in 1964 by Corey.[95] By judicious choice of experimental conditions, undesired side products were minimized and 7,7-dimethylbicyclo[4.2.0] octan-2-one obtained in 35–45% yield, the *trans* to *cis* ratio being about 4:1:

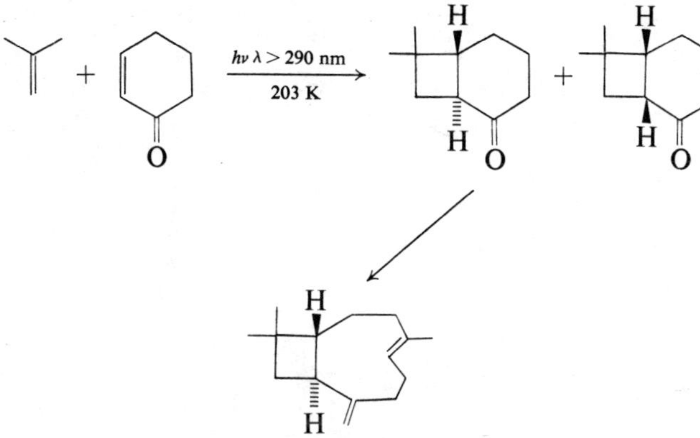


**46**

Given the right geometry, intramolecular cycloaddition may occur leading to structures often accessible only with difficulty by conventional means, for example,[96]

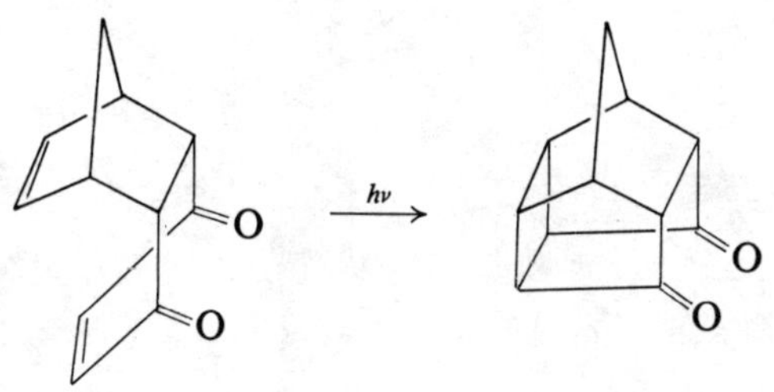


A key step in Eaton's synthesis of cubane is also a reaction of this type,[97]

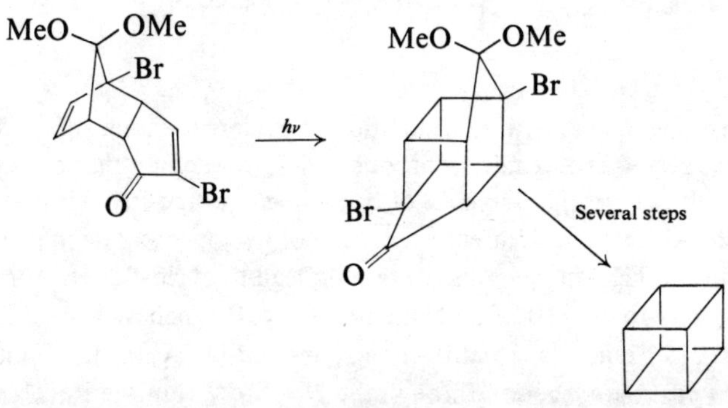

In the case of 2-cyclohexenone and substituted olefins, Corey[98] has found that olefins which are good $\pi$ donors have high reactivity and high orientational specificity. It is believed that after absorption of a photon, the ketone, probably in its $^3(n, \pi^*)$ state, forms an intermediate $\pi$ complex with the olefin which then collapses to a diradical. For positively substituted olefins, the most likely geometry is,

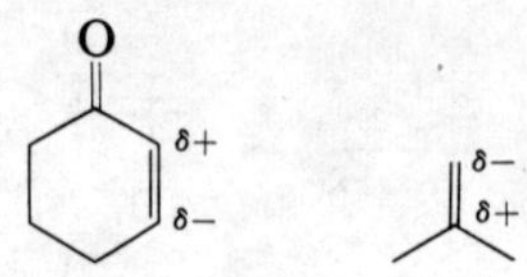

The entire reaction is then,

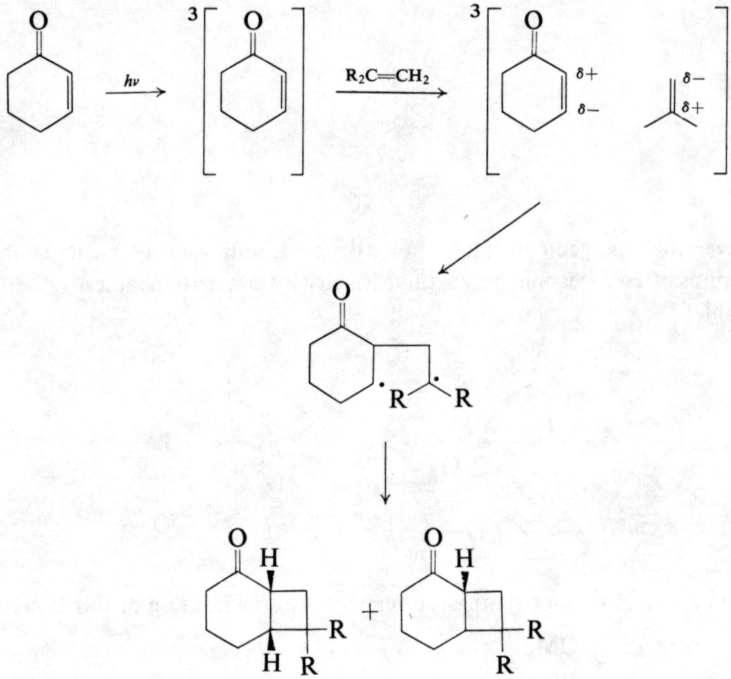

The benzophenone sensitized irradiation of cyclopentenone and cyclohexene does not give rise to cycloaddition products, despite the fact that cyclopentenone quenches both the phosphorescence of benzophenone and its photoreduction by propan-2-ol. However, both direct irradiation of cyclopentenone in propan-2-ol/ cyclohexene and high energy sensitization do lead to cycloadducts, for which a triplet energy of about 310 kJ mol$^{-1}$ is necessary. It is believed that the $T_1$ state ($E_T \simeq 255$ kJ mol$^{-1}$) is additively inactive and for cycloaddition it seems necessary to postulate reaction through the $T_2$ state[99] (but see Ref. 100).

Acetylenes will also photoadd,[101]

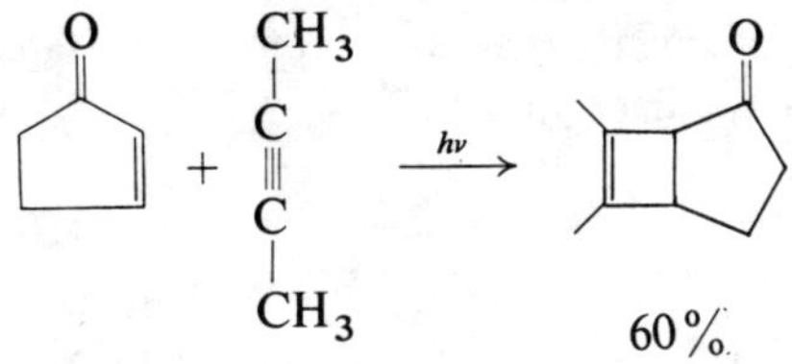

as will β-diketones,[102] reacting in their enolic form.

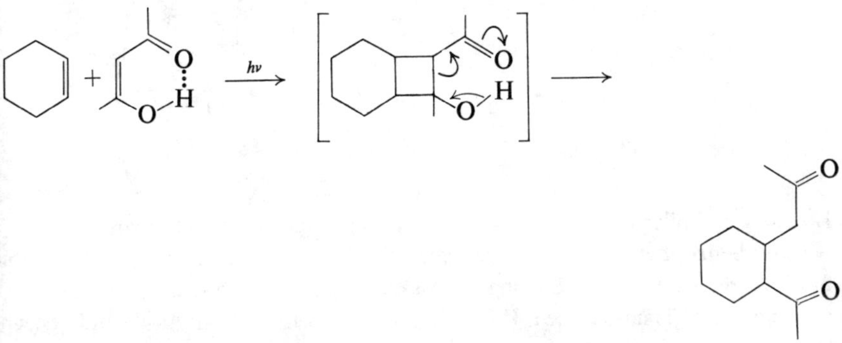

## 4.3.9 *Photocycloadditions involving benzene*

Irradiation of solutions of maleic anhydride in benzene leads to a 2:1 adduct, formally derived by 1,2-photocycloaddition, followed by thermal 1,4-addition,[103]

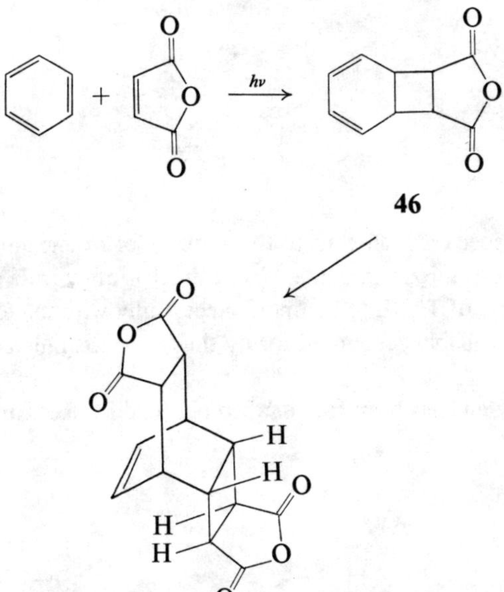

46

A similar reaction is observed between benzene and *N*-butylmaleimide.[104] Trapping experiments with tetracyanoethylene (TCNE) produces a 1:1:1 TCNE/*N*-butylmaleimide/benzene adduct, **48** thereby establishing an intermediate, **47**, having a 1,3-diene system,[42]

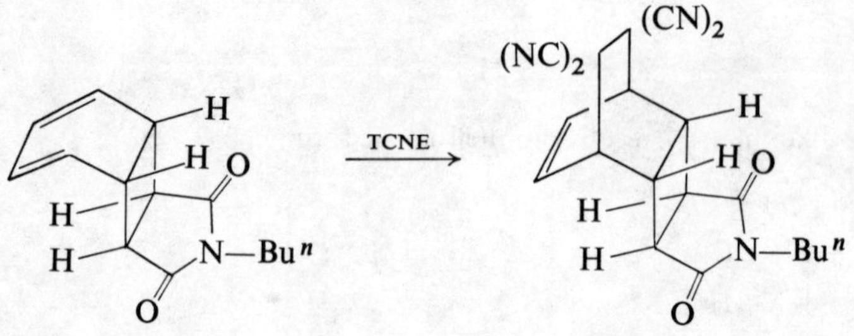


**47** **48**

However, a similar experiment with maleic anhydride and benzene with or without benzophenone as sensitizer, gives the 2:1 adduct as the only recognizabl product, recovery of TCNE being almost quantitative.[42] Evidently, a 1:1 adduct of structure **47** is not formed. It is well known that maleic anhydride and benzen form a 1:1 charge transfer complex[105] and that both its excited singlet and triplet forms are precursors of the 2:1 adduct.[106] Bryce-Smith, therefore, concludes[42] that this adduct (2:1) results from interaction of the excited CT complex with an unexcited molecule of maleic anhydride and suggests[42] that the excited species involved leads to a zwitterion, which has as one of its canonical forms,

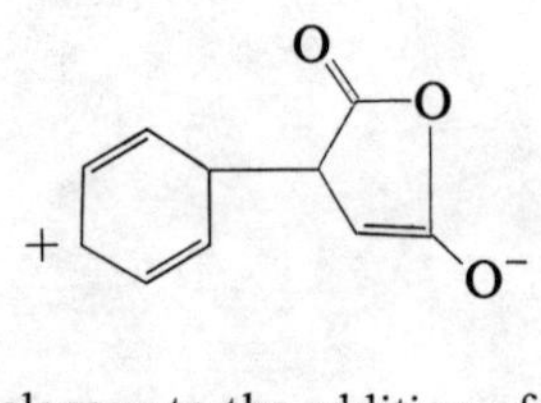


Reaction of this species, analogous to the addition of maleic anhydride to benzenonium hexafluoroantimonate,[107] with maleic anhydride gives the 2:1 adduct. The failure of TCNE to compete successfully with maleic anhydride in this last stage is probably accounted for by the fact of its much weaker nucleophilicity.[42]

Trimethylethylene has been found[108] to photoadd to benzonitrile to form a bicycloöctadiene.

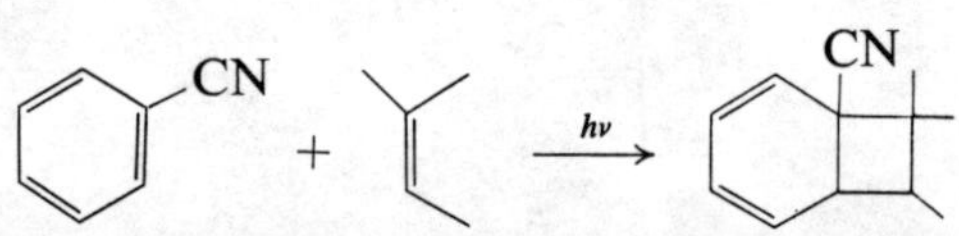


**130**

and similarly acrylonitrile photoadds to benzene,[109]

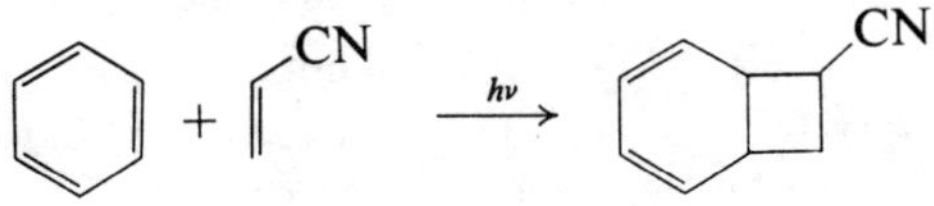

These latter observations together with the *N*-butylmaleimide–benzene cycloaddition have been interpreted[110] in terms of the allowed *cis-ortho/cis*-1,2-mode of photoreaction between benzene and ethylene-like molecules, in which either the ethylene is in its lowest singlet state (*N*-butylmaleimide and benzene), or the ethylene or aromatic component has marked acceptor properties (acrylonitrile and benzene, and benzonitrile and a mono-olefin).

A number of acetylenes, among them dimethyl acetylene dicarboxylate, also react, presumably via the intermediacy of a bicycloöctatriene, to give substituted cycloöctatetraenes,[111]

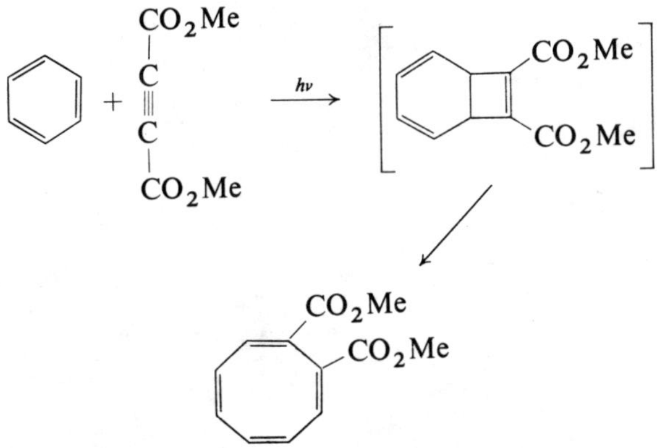

### *4.3.10 Oxetan formation—the Paterno-Büchi reaction*[112,113]

The photocycloaddition of olefins to either aldehydes or ketones produces substituted trimethylene oxides, known as oxetans. For example, irradiation of benzophenone and isobutylene leads to two isomeric products,[114]

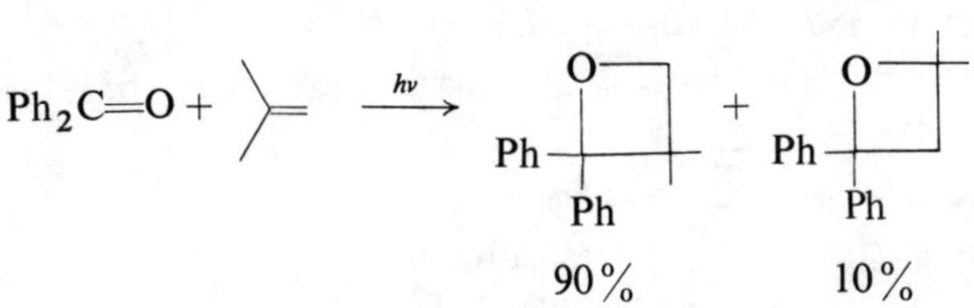

The reaction is believed to proceed in two discrete steps, thought to involve the carbonyl $^3(n, \pi^*)$ state[114] and the orientation arrived at by a consideration of the relative stabilities of the various ground state biradical intermediates.

9-Anthraldehyde shows a wavelength dependent reaction[115] when irradiated in 2-methylbut-2-ene, dimerization across the *meso*-positions occurring at wavelengths > 410 nm and oxetan formation at wavelengths < 410 nm. The mechanistic significance[116,117,118] of this is as yet uncertain but it could imply either reaction through $T_2\,(n, \pi^*)$ for $\lambda < 410$ nm or the preferential dimerization of $S_1(\pi, \pi^*)$ over intersystem crossing for $\lambda > 410$ nm. Oxetan formation may often be suppressed either by olefin dimerization or by addition of the carbonyl compound to the olefin. These side reactions are a consequence of the relative triplet energies of ketone and olefin. Consider the case of norbornene irradiated separately with benzophenone ($E_T$ = 289 kJ mol$^{-1}$), acetophenone ($E_T$ = 310 kJ mol$^{-1}$)[119] and acetone ($E_T$ > 314 kJ mol$^{-1}$),

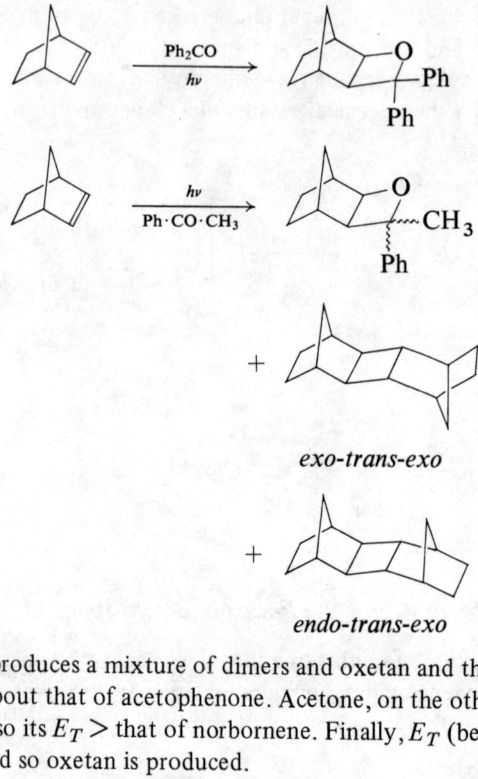


*exo-trans-exo*

*endo-trans-exo*

Acetophenone produces a mixture of dimers and oxetan and thus the $E_T$ of norbornene is about that of acetophenone. Acetone, on the other hand, gives only dimer and so its $E_T$ > that of norbornene. Finally, $E_T$ (benzophenone) < $E_T$ (norbornene) and so oxetan is produced.

The Paterno–Büchi reaction can thus be broken down into stages as follows:

(a) Excitation, $D \xrightarrow{h\nu} {}^1D$
(b) Intersystem crossing, ${}^1D \rightarrow {}^3D$
(c) Deactivation, ${}^3D$ — { (i) radiative or nonradiative; (ii) quenching; (iii) molecular rearrangement } → D
(d) Reaction, ${}^3D$ + olefin → ${}^3$[D-olefin]
D = donor (carbonyl compound).

### 4.3.11 *The Photochemical Diels–Alder reaction*

Irradiation of solutions of maleic anhydride and anthracene in dioxan leads to a 1:1-dimer, probably via the singlet state of the hydrocarbon,[120]

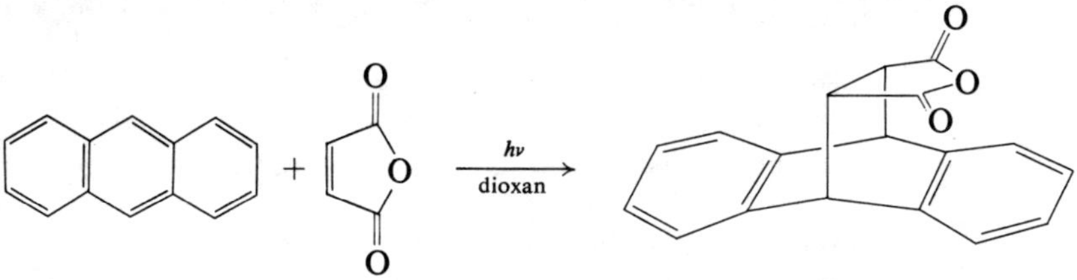


Large numbers of additions of olefins to 1,2-dicarbonyl compounds are known[121] but little mechanistic study has been undertaken,[122]

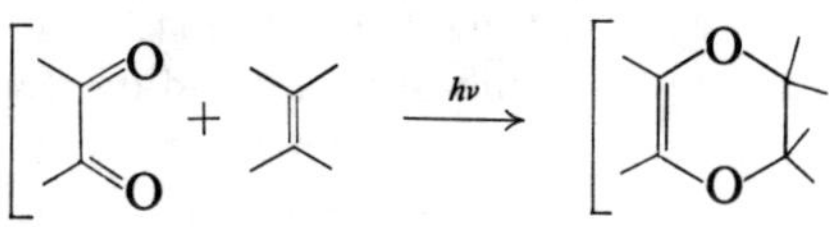


### 4.3.12 *Epidioxide formation*

The photosensitized 1,4-addition of oxygen to homoannular 1,3-dienes is a widely used and well understood reaction,[123] e.g.,

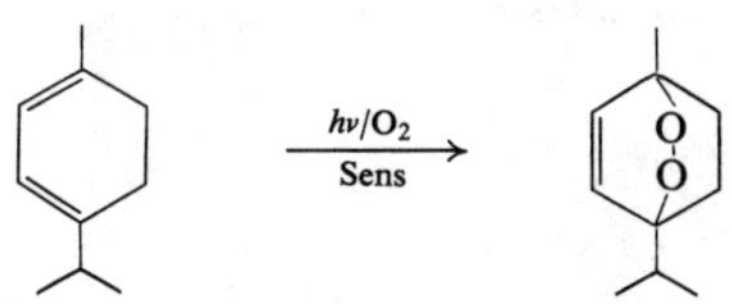


2-Terpinene Ascaridole

It has found important use in synthesis, for example, in the synthesis[124] of the plant hormone abscissic acid, **49**, in which the key step is the transformation,

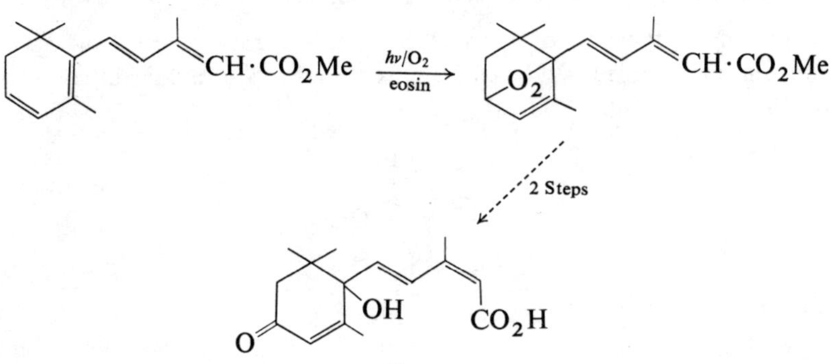


**49**

The mechanism of these photoöxidations is a subject of controversy. One

advocated by G. O. Schenck[125] involves the formation of an excited adduct between the sensitizer and oxygen,

$$\text{Sens} \xrightarrow{h\nu} \text{Sens}\,(S_1) \rightsquigarrow \text{Sens}\,(T_1)$$
$$\text{Sens}\,(T_1) + O_2(T_0) \rightarrow \dot{\text{S}}\text{ens—O—}\dot{\text{O}}$$
$$\dot{\text{S}}\text{ens—O—}\dot{\text{O}} + \text{A} \rightarrow AO_2 + \text{Sens}$$

A second mechanism invokes the intermediacy of singlet oxygen,[126, 127]

$$\text{Sens} \xrightarrow{h\nu} \text{Sens}\,(S_1) \rightsquigarrow \text{Sens}\,(T_1)$$
$$\text{Sens}\,(T_1) + O_2(T_0) \rightarrow \text{Sens}\,(S_0) + O_2(S_1)$$
$$O_2(S_1) + \text{A} \rightarrow AO_2$$

A decision between the two schemes has yet to be made and it could well be that Schenck's mechanism operates in some cases and the 'Singlet Oxygen' mechanism in others.

## 4.4 Abstraction reactions

Photochemical atom abstraction implies the removal of an atom by an electronically excited species. Often the atom concerned is hydrogen and the molecule from which it is abstracted, the solvent.

### *4.4.1 Photoreduction of ketones*

The time-honoured example is the photoreduction of benzophenone which is dealt with in section 3.2.4. There are, however, several mechanisms, discussed below, by which photoreduction may be suppressed.

(a) Alkylbenzophenones and related compounds

Flash photolysis studies[128] of *o*-benzylbenzophenone have indicated two species, one of which is regarded as the ketone in its $^3(n, \pi^*)$ state, and the other as the enol, **50**.

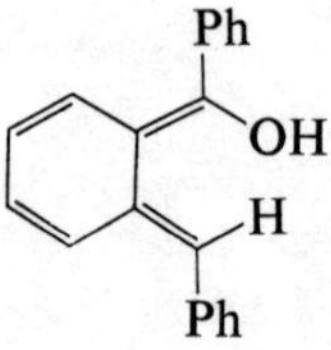


**50**

Support has been given to these assignments by the observations that irradiation of *o*-benzylbenzophenone in $CH_3OD$ leads to deuterium incorporation at the benzylic carbon,[129]

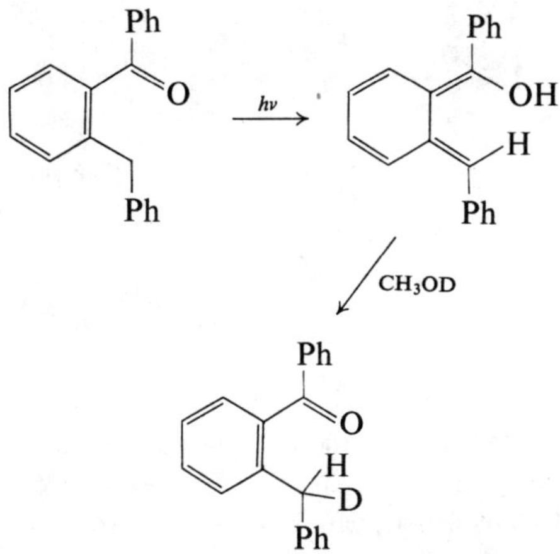

It seems most likely, therefore, that the failure of *o*-benzylbenzophenone to undergo photoreduction is a consequence of the preferential abstraction of a hydrogen atom from the *ortho* methylene group, rather than from the solvent alcohol. This is in principle an example of the Norrish Type II Process. A similar failure to undergo photoreduction is observed with *o*-hydroxy- and *o*-amino-benzophenone and here again, photoenolization has been proposed as the explanation; indeed photoprotection of plastics can be achieved by using *o*-hydroxybenzophenone as an u.v. absorber.

(b) 4-Aminobenzophenone

The failure of 4-aminobenzophenone to undergo photoreduction in isopropanol and yet to be photoreduced in cyclohexane is explained by Porter[30] in terms of the participation of an intramolecular CT state,

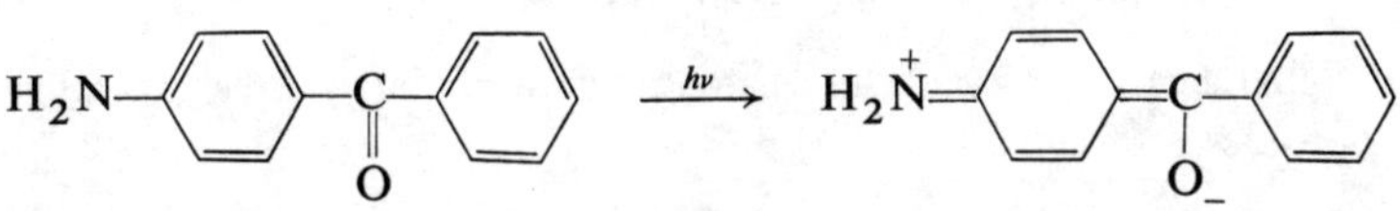

In alcoholic solvents, this state with its charge separation will be stabilized relative to the $^3(n, \pi^*)$ state of the carbonyl, the consequence being preferential excitation of the CT band. On the other hand, the ability of cyclohexene to stabilize charge separation will be considerably less, and will result in the $^3(n, \pi^*)$ state becoming the state of lowest energy, photoreduction thereby occurring. Further weight is added to this argument by the observation that 4-dimethyl-aminobenzophenone in isopropanol photoreduces only slowly, whereas photoreduction in the same solvent acidified with hydrochloric acid, proceeds substantially faster.[131] In such a medium, the nitrogen atom will be protonated rendering charge transfer separation impossible.

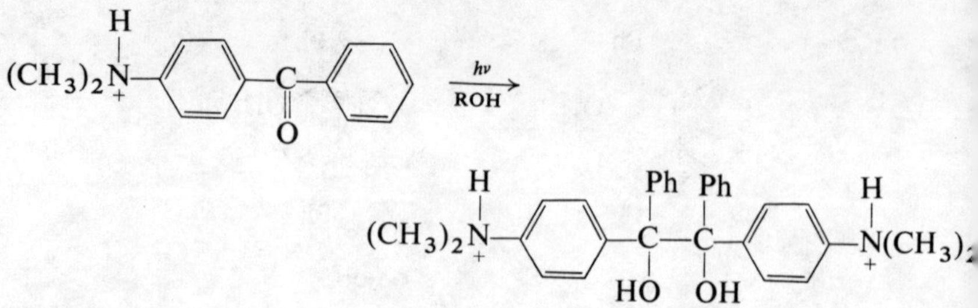

(c) 4-Hydroxybenzophenone

Particularly interesting is the fact that whereas 4-methoxybenzophenone photoreduces readily, 4-hydroxybenzophenone is inert under the same conditions. Almost certainly the explanation is that the p*K* of excited 4-hydroxybenzophenone like that of many other phenols, is much larger than in the ground state (cf. section 3.2.3). 4-Hydroxybenzophenone exists, therefore, as a phenolate anion and electron donation to the carbonyl group is strong; consequently, photoreduction is difficult. If cyclohexane is used as solvent, proton release is less, the concentration of $(n, \pi^*)$ triplets rises and photoreduction occurs.

(d) Miscellaneous

Certain aryl ketones, because of the nature of the aryl group do not have a lowest $^3(n, \pi^*)$ state and, therefore, do not photoreduce in alcoholic media. Typical examples are 2-acetonaphthone and 1-naphthaldehyde, both of which possess a lowest $^3(\pi, \pi^*)$ state. Photoreduction can, however, be brought about by use of a particularly efficient hydrogen donor, for example, tri-n-butyl stannane.[132]

## 4.4.2 *Intramolecular hydrogen abstraction*

Simple aliphatic carbonyl compounds are known to undergo three general primary processes.

(a) Norrish Type I cleavage

This process involves cleavage of the C–C bond α to the carbonyl group to form radicals and is considered in more detail on p. 74.

$$\text{R—COCH}_3 \xrightarrow{h\nu} \text{R—}\dot{\text{C}}\text{O} + \dot{\text{C}}\text{H}_3$$

(b) Norrish Type II cleavage[133]

Abstraction of a γ-hydrogen atom by an excited carbonyl group can often compete successfully with intermolecular processes and is known as the

Norrish Type II cleavage. The reaction products are normally an olefin and an enol resulting from concomitant cleavage of the α-β carbon–carbon bond.

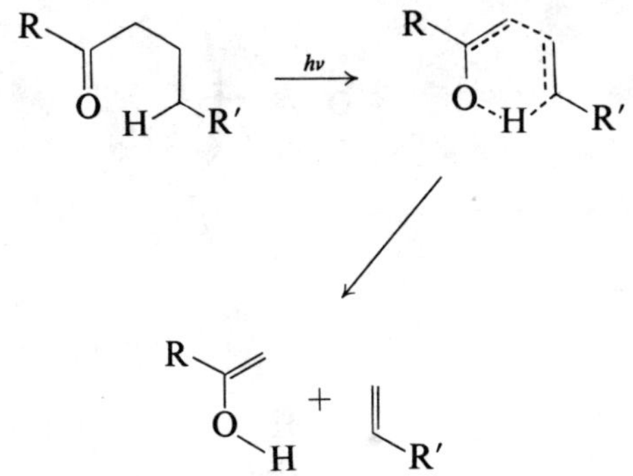


The quantum yield of photoelimination of ethylene from butyrophenone[134] at 313·0 nm and 298 K is 0·4, a figure which is highly sensitive to the electron-donating character of substituents in the aromatic rings, being 0·00 for $p$-$NH_3$, $p$-OH, and $p$-$C_6H_5$. These latter compounds have a lowest $^3(\pi, \pi^*)$ state strongly suggesting that the $^3(n, \pi^*)$ is reactive in photoabstraction.

From quencher studies it appears that aliphatic ketones undergo this reaction from either the singlet or triplet state.

(c) Cyclobutanol formation

Cyclobutanols may also be formed photochemically from ketones possessing a γ-hydrogen and Yang[135] has suggested a stepwise mechanism since irradiation of hept-6-en-2-one gives rise to both methylvinylcyclobutanols, **51**, and methylcyclohexenol, **52**.

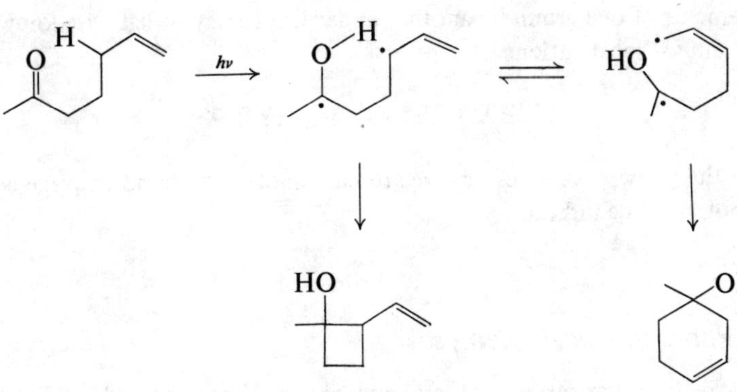


**51** **52**

However, other workers[136] have found this unconvincing and prefer a concerted mechanism since (6S)-(+)-2,6-dimethyloct-7-en-3-one, **53**, leads to (4S)-(+)-

terpinen-4-ol, **54**, among the irradiation products, arising, it is argued, by an intramolecular ene-synthesis,

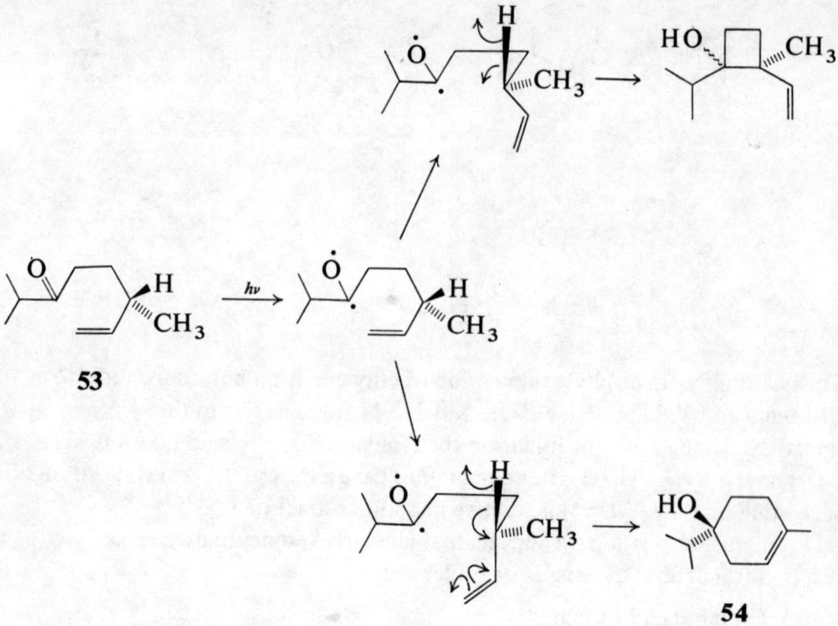

## 4.5 Organic photosubstitutions

Replacements of one group by another under the agency of light are known as photochemical substitutions:

$$RX + Y \xrightarrow{h\nu} RY + X$$

Most of the known examples involve aromatic molecules, bond cleavage being either polar or free radical.

### *4.5.1 Photochemical solvolyses*

The observation that certain isomeric nitrophenyl phosphate and sulphate esters undergo photochemical hydrolysis was made by Havinga[137] in 1956 and subsequently investigated[138] by him and Zimmerman independently. These compounds were found to be quite stable over a wide pH range and yet on illumination they cleaved to give the nitrophenol and the corresponding inorganic acid. Of particular interest was the observation that it was the *meta*-substituted compounds which underwent the most efficient hydrolysis. Irrespective of the molecularity of the reaction, by analogy with ground state chemistry, stabilization of an incipient phenolate anion by a *p*-nitro group is understandable, whereas, stabilization by a *m*-nitro group is not. This is in accordance with the selective transmission of electronic effects from phenoxy oxygen to a *para*-electron withdrawing group. The electron densities for benzene, separately monosubstituted with electron withdrawing (W) and electron donating (D) groups, in the ground and first excited states, are shown[139] in Fig. 4.33.

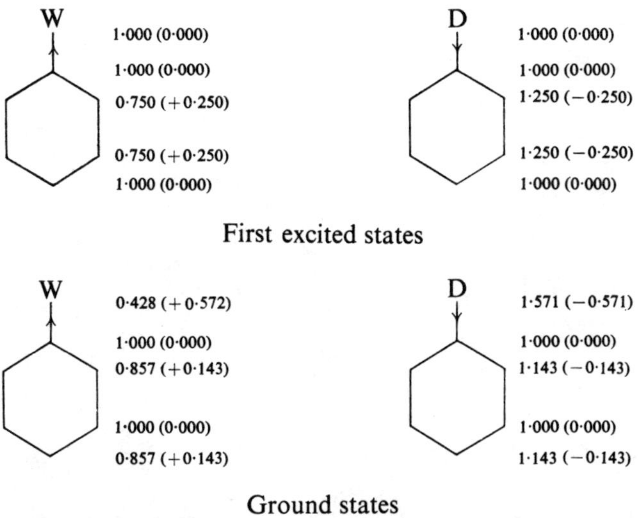


**Fig. 4.33** Numbers in parentheses are formal charges; those unparenthesized are $\pi$-electron densities.

These figures are the result of calculations based upon the simplest examples of $C_6H_5$-W and $C_6H_5$-D, in which W is $CH_2^+$ and D is $CH_2^-$, i.e., the electron densities in the benzyl cation and anion respectively. In the former case, inspection reveals that in sharp contrast to the ground state, in which electron density is reduced at the *ortho*- and *para*-positions, in the first excited state, this selective reduction is from the *ortho*- and *meta*-positions. In the latter case, the electron density in the first excited state is increased at the *ortho*- and *meta*-positions relative to the ground state. Replacement of $-CH_2^+$ by $-NO_2$ and $-CH_2^-$ by $CH_3O-$ leads to the same qualitative conclusions. An understanding of the nature of this *meta*-transmission may be obtained from Fig. 4.34. Here the seven molecular orbitals of the $\pi$ framework of the benzyl cation and the benzyl anion are represented.

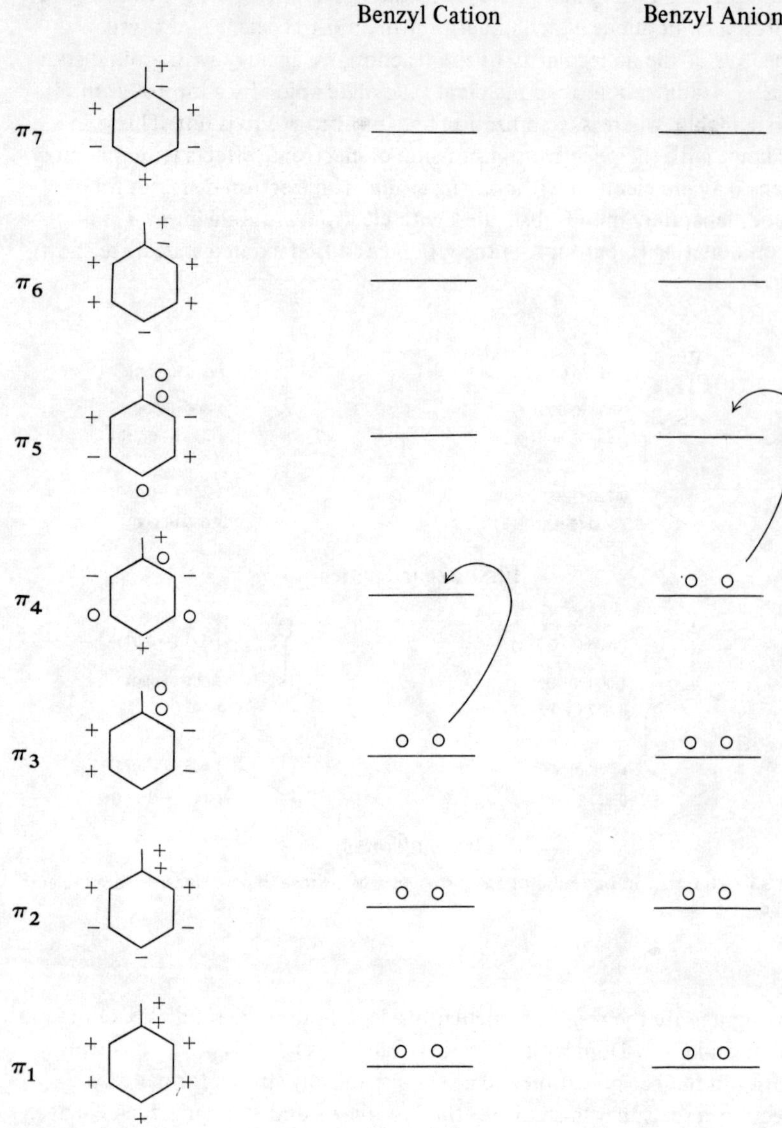


Fig. 4.34

For the benzyl cation, excitation of one electron from the highest occupied MO, in which it has density on the *meta*-position, to the lowest unoccupied MO (i.e., the electronic transition requiring minimum energy), in which it has no density on the *meta*-position, causes diminution of electron density at the *meta*-position.

The results of the electron density diagrams can be expressed in valence bond terms as,

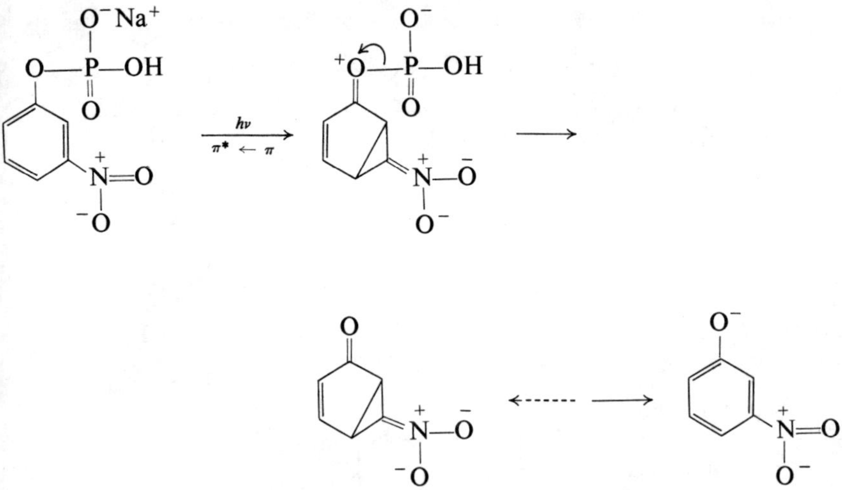

This type of behaviour is also exhibited by nitrophenyl trityl esters and cyanotrityl esters.

Inspection of the MO diagram for the benzyl anion reveals that upon excitation a situation of selective electron donation to the *meta*-position should arise and on irradiation anisole should be capable of stabilizing a positive charge at its *meta*-position. Accordingly, *m*-methoxybenzyl acetate readily undergoes photolysis in aqueous dioxan leading to the corresponding alcohol,

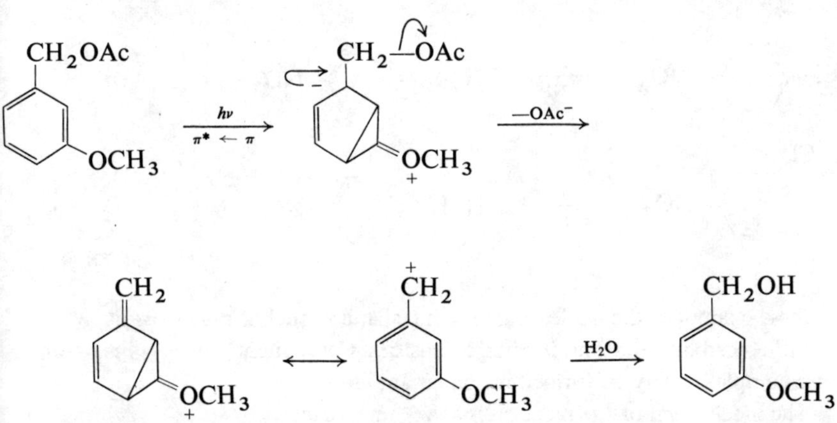

*p*-Methoxybenzyl acetate affords essentially the free radical products expected from homolytic generation of 4-methoxybenzyl and acetoxyl radicals,

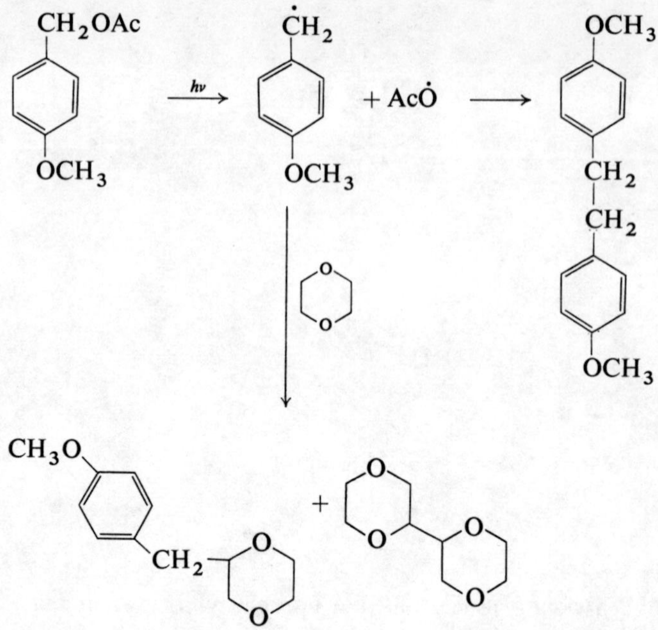

Just as in the above examples the pattern of reactivity is determined by the charge distribution in the excited state, so other reactions are known, not observed in the dark, in which the orientation follows the familiar classical pattern associated with ground state chemistry, e.g., the photoaminations of nitrobenzene,[138]

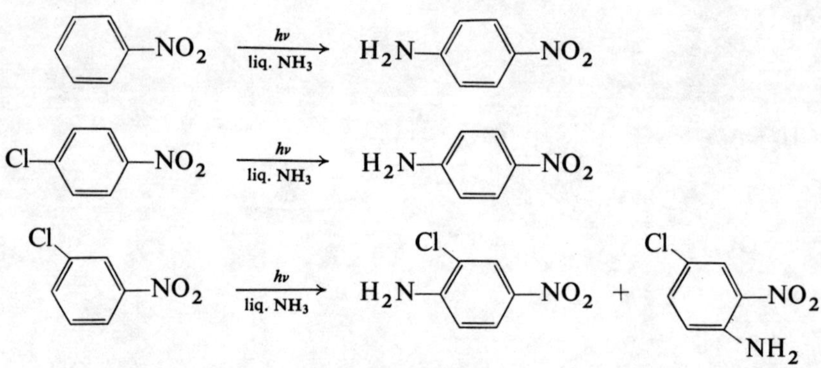

In these examples, the nucleophile is neutral and as such is much less likely to be influenced by variations in charge densities. Consequently *ortho-*, *meta-*, and *para*-complexes may be formed in similar amounts.

The mechanism of the reaction is as yet uncertain.

### 4.5.2 *Halogen substitutions*

The replacement of bromine from bromobenzene by chlorine was first reported[140] in 1903 and has been studied by Walling[141] who was able to show that the reaction is a radical chain process,

$$Cl_2 \xrightleftharpoons{h\nu} 2\dot{Cl}$$
$$\dot{Cl} + PhBr \rightarrow PhCl + \dot{Br}$$
$$\dot{Br} + Cl_2 \rightarrow BrCl + \dot{Cl}$$
$$2BrCl \rightarrow Br_2 + Cl_2$$

The critical step in the reaction, if a direct displacement on carbon, could either be a concerted one step process, a two step process involving a relatively stable $\sigma$-complex, or some intermediate between these two extremes. There is the third possibility that the reaction proceeds by rearrangement of an initially loose $\pi$-complex,

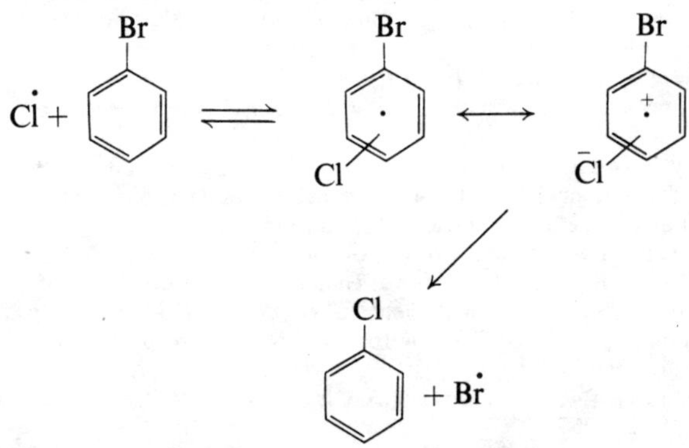


The effects of substituents have been reported[142] and this last mechanism is certainly in keeping with the decreased reactivity of bromobenzenes with electron-withdrawing groups.

## References

1. R. B. Woodward and R. Hoffmann, *Ang. Chem. Int. Ed. Eng.*, **8**, 781 (1969).
2. W. A. Noyes, Jr., G. B. Porter, and J. E. Jolley, *Chem. Rev.*, **56**, 49 (1956).
3. R. Srinivasan, *Adv. Photochem.*, **1**, 83 (1963).
4. R. E. Rebbert and P. Ausloos, *J. Phys. Chem.*, **66**, 2253 (1962).
5. H. M. Frey, *Adv. Photochem.*, **4**, 225 (1966).
6. H. M. Frey and I. D. R. Stevens, *J. Chem. Soc.*, 1700 (1965).
7. T. V. Van Auken and K. L. Rinehart, *J. Amer. Chem. Soc.*, **84**, 3736 (1962).
8. P. Dowd, *J. Amer. Chem. Soc.*, **88**, 2587 (1966).
9. G. Herzberg and J. Shoosmith, *Nature* (Lond.), **183**, 1801 (1959).
10. F. A. L. Anet, R. F. W. Bader, and A.-M. Van der Auwera, *J. Amer. Chem. Soc.*, **82**, 3217 (1960).
11. W. Kirmse, *Carbene Chemistry*, Academic Press, New York, p. 34 (1964).
12. S. J. Rhoads, in *Molecular Rearrangements*, Ed., P. de Mayo, Interscience, New York, p. 702 (1963).

13. K. Hafner, W. Kaiser, and R. Puttner, *Tetrahedron Letters*, 3953 (1964).
14. D. H. R. Barton and L. R. Morgan, Jr., *J. Chem. Soc.*, 622 (1962).
15. D. H. R. Barton and A. N. Starrat, *J. Chem. Soc.*, 2444 (1965).
16. R. M. Moriarty and M. Rahman, *Tetrahedron*, **21**, 2877 (1965).
17. A. L. Nussbaum and C. H. Robinson, *Tetrahedron*, **17**, 351 (1962).
18. P. Kabasakalian and E. R. Townley, *J. Amer. Chem. Soc.*, **84**, 2711, 2716, 2718, 2723, 2724 (1962).
19. D. H. R. Barton, J. M. Beaton, L. E. Geller, and M. M. Pechet, *J. Amer. Chem. Soc.*, **82**, 2640 (1960).
20. M. Akhtar, *Adv. Photochem.*, **2**, 263 (1964).
21. D. H. R. Barton and J. M. Beaton, *J. Amer. Chem. Soc.*, **82**, 2641 (1960).
22. D. H. R. Barton, H. P. Faro, E. P. Serebryakov, and N. F. Woolsey, *J. Chem. Soc.*, 2438 (1965).
23. J. Saltiel and L. Matts, *J. Amer. Chem. Soc.*, **89**, 2232 (1967).
24. R. S. Mulliken and C. C. J. Roothan, *Chem. Rev.*, **41**, 219 (1947).
25. R. B. Cundall, *Prog. Reaction Kinetics*, **2**, 166 (1964).
26. P. J. Wagner and G. S. Hammond, *Adv. Photochem.*, **5**, 21 (1968).
27. Z. R. Grabowski and A. Bylina, *Trans. Faraday Soc.*, **60**, 1131 (1964).
28. G. S. Hammond, J. Saltiel, A. A. Lamola, N. J. Turro, J. S. Bradshaw, D. O. Cowan, R. C. Counsell, V. Vogt, and C. Dalton, *J. Amer. Chem. Soc.*, **86**, 3197 (1964).
29. L. B. Jones and G. S. Hammond, *J. Amer. Chem. Soc.*, **87**, 4219 (1965).
30. O. L. Chapman, *Adv. Photochem.*, **1**, 323 (1963).
31. R. E. K. Winter, *Tetrahedron Letters*, 1207 (1965).
32. K. M. Shumate, P. N. Newman, and G. J. Fonken, *J. Amer. Chem. Soc.*, **87**, 3996 (1965).
33. E. N. Marvell, G. Caple, and B. Schlatz, *Tetrahedron Letters*, 385 (1965).
34. E. Vogel, W. Grimme, and E. Dinne, *Tetrahedron Letters*, 391 (1965).
35. H. Sponer and E. Teller, *Phys. Rev.*, **13**, 75 (1941).
36. E. N. Marvell and J. Seubert, *J. Amer. Chem. Soc.*, **89**, 3377 (1967).
37. R. F. C. Brown, R. C. Cookson, and J. Hudec, *Tetrahedron*, **24**, 3955 (1968).
38. J. A. Berson and G. L. Nelson, *J. Amer. Chem. Soc.*, **89**, 5303 (1967).
39. G. S. Hammond, P. Wyatt, C. D. DeBoer, and N. J. Turro, *J. Amer. Chem. Soc.*, **86**, 2532 (1964).
40. H. E. Zimmerman, R. W. Binkley, R. S. Givens, and M. A. Sherwin, *J. Amer. Chem. Soc.*, **89**, 3932 (1967).
41. H. E. Zimmerman and P. S. Mariano, *J. Amer. Chem. Soc.*, **91**, 1718 (1969).
42. D. Bryce-Smith, *Pure Appl. Chem.*, **16**, 47 (1968).
43. A. E. Douglas, *Disc. Faraday Soc.*, **35**, 158 (1963).
44. H. J. F. Angus, J. M. Blair, and D. Bryce-Smith, *J. Chem. Soc.*, 2003 (1960).
45. L. Kaplan and K. E. Wilzbach, *J. Amer. Chem. Soc.*, **89**, 1030 (1967).
46. K. E. Wilzbach and L. Kaplan, *J. Amer. Chem. Soc.*, **87**, 4004 (1965).
47. K. E. Wilzbach and L. Kaplan, *J. Amer. Chem. Soc.*, **88**, 2066 (1966).
48. D. Bryce-Smith, A. Gilbert, and B. H. Orger, *Chem. Commun.*, 512 (1966).
49. K. Koltzenburg and K. Kraft, *Tetrahedron Letters*, 389 (1966).
50. I. Haller, *J. Chem. Phys.*, **47**, 1117 (1967).
51. M. S. De Groot and J. H. van der Waals, *Mol. Phys.*, **6**, 545 (1963).
52. E. E. Van Tamelen and S. P. Pappas, *J. Amer. Chem. Soc.*, **85**, 3297 (1963).
53. E. E. Van Tamelen, *Ang. Chem. Int. Ed. Eng.*, **4**, 738 (1965).
54. E. Farenhorst, *Tetrahedron Letters*, 6465 (1966).
55. P. J. Kropp, *Org. Photochem.*, **1**, 1 (1967).
56. H. E. Zimmerman, *Adv. Photochem.*, **1**, 183 (1963).
57. W. G. Dauben, K. Koch, O. L. Chapman, and S. L. Smith, *J. Amer. Chem. Soc.*, **83**, 1768 (1961).
58. V. I. Stenberg, *Org. Photochem.*, **1**, 127 (1967).
59. M. R. Sandner, E. Hedaya, and D. J. Trecker, *J. Amer. Chem. Soc.*, **90**, 7249 (1968).
60. J. W. Meyer and G. S. Hammond, *J. Amer. Chem. Soc.*, **92**, 2187 (1970).
61. M. S. Kharasch, G. Stampa, and W. Nudenberg, *Science*, **116**, 309 (1952).
62. F. B. Mallory, C. S. Wood, and J. T. Gordon, *J. Amer. Chem. Soc.*, **86**, 3094 (1964).

63. K. A. Muszkat, D. Gegiou, and E. Fischer, *Chem. Commun.*, 447 (1965).
64. G. B. Badger, R. J. Drewer, and G. E. Lewis, *Aust. J. Chem.*, **17**, 1036 (1964).
65. D. Elad, *Org. Photochem.*, **2**, 168 (1969).
66. S. Y. Wang, *Federation Proc., Suppl. No. 15*, **24**, 71 (1965).
67. W. H. Urry, F. W. Stacey, E. S. Huyser, and O. O. Juveland, *J. Amer. Chem. Soc.*, **76**, 450 (1954).
68. R. Stoermer and H. Stockman, *Chem. Ber.*, **48**, 1786 (1914).
69. J. A. Marshall, *Acc. Chem. Res.*, **2**, 33 (1969).
70. W. H. Urry and O. O. Juveland, *J. Amer. Chem. Soc.*, **80**, 3322 (1958).
71. D. Elad and J. Rokach, *J. Org. Chem.*, **29**, 1855 (1964).
72. M. S. Karasch, H. W. Urry, and B. M. Kuderna, *J. Org. Chem.*, **14**, 248 (1949).
73. P. de Mayo, J. B. Stothers, and W. Templeton, *Canad. J. Chem.*, **39**, 488 (1961).
74. H. D. Scharf and F. Korte, *Chem. Ber.*, **97**, 2425 (1964).
75. D. R. Arnold, D. J. Trecker, and E. B. Whipple, *J. Amer. Chem. Soc.*, **87**, 2596 (1965).
76. G. S. Hammond, N. J. Turro, and R. H. S. Liu, *J. Amer. Chem. Soc.*, **28**, 3297 (1963).
77. R. H. S. Liu, N. J. Turro, and G. S. Hammond, *J. Amer. Chem. Soc.*, **87**, 3406 (1965).
78. D. Valentine, N. J. Turro, and G. S. Hammond, *J. Amer. Chem. Soc.*, **86**, 5202 (1964).
79. R. H. S. Liu and G. Hammond, *J. Amer. Chem. Soc.*, **86**, 1892 (1964).
80. K. J. Crowley, *Proc. Chem. Soc.*, 334 (1962).
81. J. Fritzsche, *J. prakt. Chemie*, **101**, 333 (1867).
82. R. Calas, R. Lalande, J.-G. Faugère, and F. Moulines, *Bull. Soc. Chim. France*, 119 (1965).
83. E. J. Bowen, *Adv. Photochem.*, **1**, 23 (1963).
84. J. S. Bradshaw and G. S. Hammond, *J. Amer. Chem. Soc.*, **85**, 3953 (1963).
85. G. S. Hammond, C. A. Stout, and A. A. Lamola, *J. Amer. Chem. Soc.*, **86**, 3103 (1964).
86. E. J. Corey and J. Streith, *J. Amer. Chem. Soc.*, **86**, 950 (1964).
87. A. Padwa and R. Hartman, *J. Amer. Chem. Soc.*, **86**, 4212 (1964).
88. E. C. Taylor and R. O. Kan, *J. Amer. Chem. Soc.*, **85**, 776 (1963).
89. M. D. Cohen and G. M. J. Schmidt, *J. Chem. Soc.*, 1996 (1964).
90. M. D. Cohen, G. M. J. Schmidt, and F. I. Sonntag, *J. Chem. Soc.*, 2000 (1964).
91. J. Bergman, K. Osaki, G. M. J. Schmidt, and F. I. Sonntag, *J. Chem. Soc.*, 2021 (1964).
92. R. C. Cookson, D. A. Cox, and J. Hudec, *J. Chem. Soc.*, 4499 (1961).
93. A. Cox, P. de Mayo, and R. W. Yip, *J. Amer. Chem. Soc.*, **88**, 1043 (1966).
94. R. Robson, P. W. Grubb, and J. A. Barltrop, *J. Chem. Soc.*, 2153 (1964).
95. E. J. Corey, R. B. Mitra, and H. Uda, *J. Amer. Chem. Soc.*, **86**, 485 (1964).
96. R. C. Cookson, E. Crundwell, R. R. Hill, and J. Hudec, *J. Chem. Soc.*, 3062 (1964).
97. P. Eaton and T. W. Cole, *J. Amer. Chem. Soc.*, **86**, 3157 (1964).
98. E. J. Corey, J. D. Bass, R. LeMahieu, and R. B. Mitra, *J. Amer. Chem. Soc.*, **86**, 5570 (1964).
99. P. de Mayo, J.-P. Pete, and M. Tchir, *J. Amer. Chem. Soc.*, **89**, 5712 (1967).
100. P. de Mayo, A. A. Nicholson, and M. F. Tchir, *Canad. J. Chem.*, **48**, 225 (1970).
101. P. E. Eaton, *Tetrahedron Letters*, 3695 (1964).
102. P. de Mayo and H. Takeshita, *Canad. J. Chem.*, **41**, 440 (1963).
103. H. J. F. Angus and D. Bryce-Smith, *J. Chem. Soc.*, 4791 (1960).
104. D. Bryce-Smith and M. A. Hems, *Tetrahedron Letters*, 1895 (1966).
105. L. J. Andrews and R. M. Keefer, *J. Amer. Chem. Soc.*, **75**, 3776 (1953).
106. D. Bryce-Smith and J. E. Lodge, *J. Chem. Soc.*, 2675 (1962).
107. G. Olah, *J. Amer. Chem. Soc.*, **87**, 1103 (1965).
108. J. G. Atkinson, D. E. Ayer, G. Büchi, and E. W. Robb, *J. Amer. Chem. Soc.*, **85**, 2257 (1963).
109. B. E. Job and J. D. Littlehailes, *J. Chem. Soc.* (*C*), 886 (1968).
110. D. Bryce-Smith, *Chem. Commun.*, 806 (1969).
111. D. Bryce-Smith and J. E. Lodge, *J. Chem. Soc.*, 695 (1963).
112. O. L. Chapman and G. Lenz, *Org. Photochem.*, **1**, 283 (1967).

113. D. R. Arnold, *Adv. Photochem.*, **6**, 301 (1968).
114. D. R. Arnold, R. L. Hinman, and A. H. Glick, *Tetrahedron Letters*, 1425 (1964).
115. N. C. Yang, M. Nussim, M. J. Jorgenson, and S. Murov, *Tetrahedron Letters*, 3657 (1964).
116. N. C. Yang, R. L. Loeschen, and D. Mitchell, *J. Amer. Chem. Soc.*, **89**, 5466 (1967).
117. N. C. Yang and R. L. Loeschen, *Tetrahedron Letters*, 2571 (1968).
118. N. J. Turro, in *Technique of Organic Chemistry*, **XIV**, Ed., A. Weissberger, Interscience, New York, p. 133 (1969).
119. D. R. Arnold, D. J. Trecker, and E. B. Whipple, *J. Amer. Chem. Soc.*, **87**, 2596 (1965).
120. J. P. Simons, *Trans. Faraday Soc.*, **56**, 391 (1960).
121. A. Schönberg, G. O. Schenck, and O.-A. Neumüller, *Preparative Organic Photochemistry*, Springer-Verlag, Berlin, p. 119 (1968).
122. C. H. Krauch, S. Farid, and G. O. Schenck, *Chem. Ber.*, **98**, 3102 (1965).
123. G. O. Schenck and K. Ziegler, *Naturwiss.*, **32**, 157 (1945).
124. J. W. Cornforth, R. Mallaby, and G. Ryback, *J. Chem. Soc. (C)*, 1565 (1968).
125. K. Gollnick and G. O. Schenck, *Pure Appl. Chem.*, **9**, 507 (1964).
126. H. Kautsky and H. de Bruijn, *Naturwiss.*, **19**, 1043 (1931).
127. C. S. Foote and S. Wexler, *J. Amer. Chem. Soc.*, **86**, 3880 (1964).
128. E. F. Zwicker, L. I. Grossweiner, and N. C. Yang, *J. Amer. Chem. Soc.*, **85**, 2671 (1963).
129. N. C. Yang and C. Rivas, *J. Amer. Chem. Soc.*, **83**, 2213 (1961).
130. G. Porter and P. Suppan, *Proc. Chem. Soc.*, 191 (1964).
131. S. G. Cohen and M. N. Siddiqui, *J. Amer. Chem. Soc.*, **86**, 5047 (1964).
132. G. S. Hammond and P. A. Leermakers, *J. Amer. Chem. Soc.*, **84**, 207 (1962).
133. For a discussion see *Photochemistry* by J. G. Calvert and J. N. Pitts, Jr., John Wiley and Sons, Inc., New York, p. 382 (1966).
134. E. J. Baum, J. K. S. Wan, and J. N. Pitts, *J. Amer. Chem. Soc.*, **88**, 2652 (1966).
135. N. C. Yang, A. Morduchowitz, and D.-D. H. Yang, *J. Amer. Chem. Soc.*, **85**, 1017 (1963).
136. K. H. Schulte-Elte and G. Ohloff, *Tetrahedron Letters*, 1143 (1964).
137. E. Havinga, R. O. de Jongh, and W. Dorst, *Rec. Trav. chim.*, **75**, 378 (1956).
138. For a review of this and other reactions see E. Havinga, R. O. de Jongh, and M. E. Kronenberg, *Helv. Chim. Acta*, **50**, 2550 (1967).
139. H. E. Zimmerman and V. R. Sandel, *J. Amer. Chem. Soc.*, **85**, 915 (1963).
140. A. Eibner, *Chem. Ber.*, **36**, 1229 (1903).
141. B. Miller and C. Walling, *J. Amer. Chem. Soc.*, **79**, 4187 (1957).
142. B. Milligan, R. L. Bradow, J. E. Rose, H. E. Hubbert, and A. Roe, *J. Amer. Chem. Soc.*, **84**, 158 (1962).

# 5. Experimental procedures

## 5.1 Light sources

### 5.1.1 Lamps

The characteristics of light sources are of major concern to photochemists. In much of the early work, especially of Ciamician and Silber,[1] sunlight was of necessity the source of radiant energy used in bringing about photochemical transformations. Such a situation, of course, no longer prevails and today mercury arc lamps are usually the source of choice.

Lamps operating at $10^{-6}$ bar and near room temperature are low pressure mercury arcs. Their emission spectrum comprises two bands centred at 253·7 nm and 184·9 nm (cf. section 2.1), corresponding respectively to the transitions,

$$Hg(^3P_1) \rightarrow Hg(^1S_0) + h\nu$$

$$Hg(^1P_1) \rightarrow Hg(^1S_0) + h\nu$$

However, unless a Suprasil quartz envelope is used, this second line will not be transmitted. These lines are *resonance* lines, by which is meant that they are both the longest wavelengths capable of inducing fluorescence between the two pairs of states involved. Since the 253·7 nm and 184·9 nm emissions are *reversed* lines (see below) only low pressure lamps are useful in mercury photosensitization processes (section 2.1). Low pressure lamps are essentially free from continuum and consequently are often employed as sources of monochromatic radiation of wavelength 253·7 nm.

Medium pressure mercury arc lamps operate at about 1 bar (and higher pressures). The centres of the 253·7 nm and 184·9 nm bands are both missing since absorption of these resonance radiations takes place by the relatively cool atoms near the glass envelope. This situation is shown in Fig. 5.1.

The 253·7 nm and 184·9 nm lines are thus known as *reversed* radiations which, although useful in initiating direct photochemical reaction, are incapable of functioning as sources of radiation for mercury sensitization processes when generated under these conditions.

In addition to the 253·7 nm and 184·9 nm lines mentioned above, medium pressure lamps display a large number of other lines. These arise from bombardment of the $^3P_1$ atoms by species such as electrons and molecules, resulting partly in their excitation to higher states and partly in their deactivation to the $^3P_0$ state, emission from which leads to 265·4 nm radiation (section 2.1). Other useful lines at about 313·0 nm and 365·0 are present (cf. Fig. 2.1). The

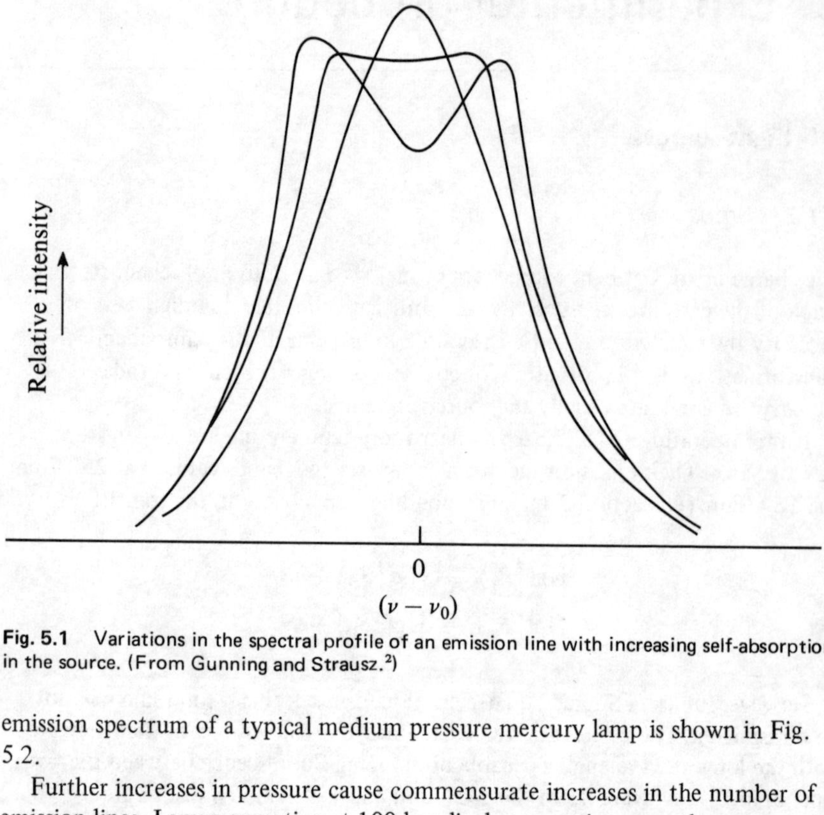


**Fig. 5.1** Variations in the spectral profile of an emission line with increasing self-absorption in the source. (From Gunning and Strausz.[2])

emission spectrum of a typical medium pressure mercury lamp is shown in Fig. 5.2.

Further increases in pressure cause commensurate increases in the number of emission lines. Lamps operating at 100 bar display a continuum and are particularly useful where high intensity sources are required. Such high pressure mercury arc lamps are, however, subject to intensity fluctuations with time and for much quantitative work, especially where it is necessary to use a collimated beam, point sources are more desirable. These lamps depend upon discharge in mercury or mercury/xenon, operate in the region of 15 bar and usually possess three electrodes, the third being the starting electrode. The emission spectrum of a typical point source mercury lamp, in this case an Osram HBO 200 W, is shown in Fig. 5.3.

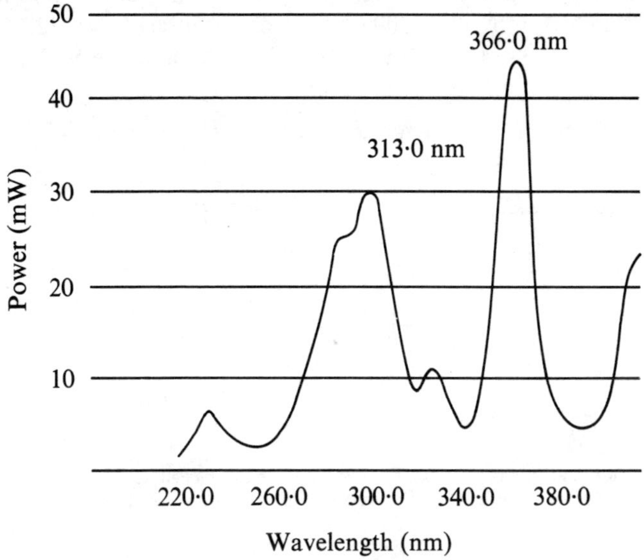


**Fig. 5.2** Emission spectrum of a typical medium pressure mercury arc lamp.

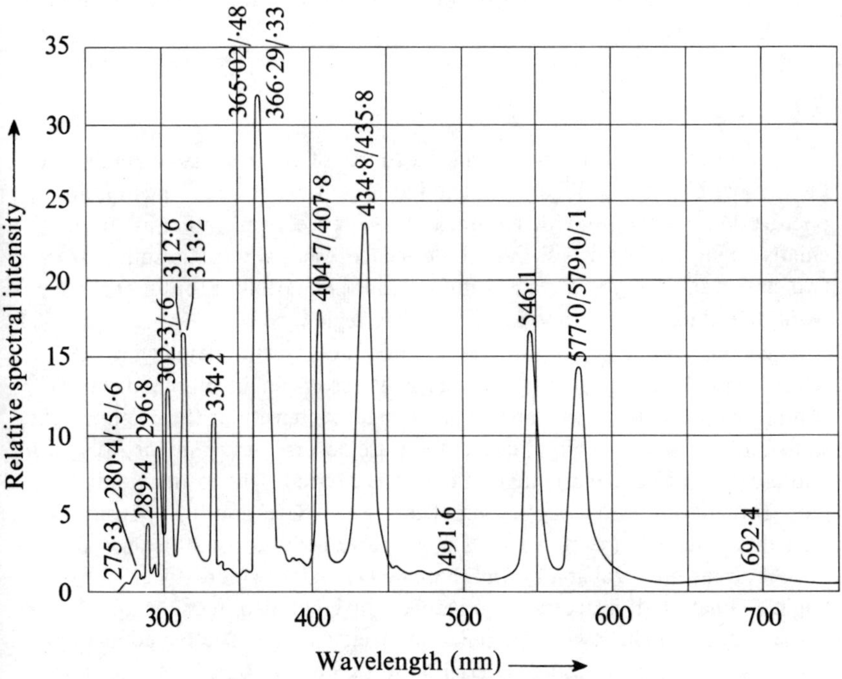


**Fig. 5.3** Relative spectral intensity distribution of HBO 200. (Reproduced by kind permission of Osram G.m.b.H.)

Discharge lamps available for studies in the far u.v. include the low pressure discharge from hydrogen (80·0–200·0 nm), from krypton (116·5, 123·6 nm) and from xenon (129·6, 147·0 nm) (Fig. 5.4) while radiation in the 50·0–165·0 nm

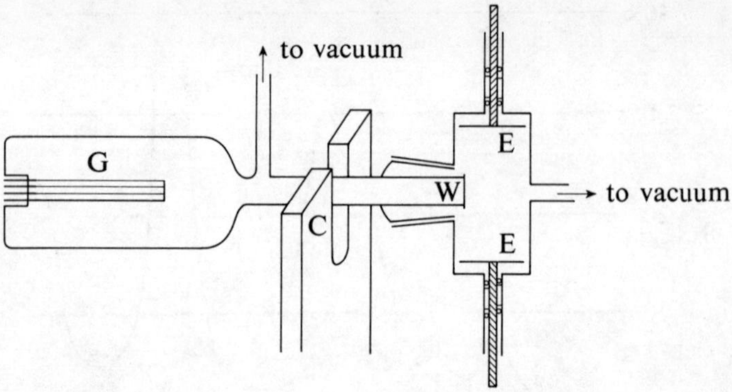


**Fig. 5.4** A rare gas resonance lamp attached to a photolysis cell. G, titanium getter wires in the 1000 ml ballast volume; C, microwave cavity; W, window; E, nickel electrodes. From P. Ausloos.[3]

region is provided by low pressure discharges using hydrogen, helium, or argon in conjunction with a hot cathode.

Fluorescent tubes are also used in photochemical work as are incandescent bulbs. A useful compilation of market data on various sources has been made by Schenck.[4]

### *5.1.2 Lasers*

The advent of lasers in 1960[5] provided a source of electromagnetic radiation in the u.v., visible, and i.r. regions unique in the respects of directionality, power, polarization, monochromatic nature, and coherence. Application has been chiefly to physical studies such as fluorescence, especially that resulting from two- and even three-photon absorption, and flash photolysis, but preliminary work on ordinary photochemical effects has begun.

The name is an acronym for *l*ight *a*mplification by *s*timulated *e*mission of *r*adiation. Whilst the rate of ordinary light absorption by a ground-state atom or molecule is proportional to the incident intensity of the exciting light, emission from the corresponding excited state occurs both spontaneously, as in fluorescence, and under stimulation from the exciting light to an extent proportional to both the population of the excited level and the intensity of the stimulating light (section 1.4). The proportionality constant is termed the Einstein transition probability for stimulated emission. To greatly amplify the original radiation, it is necessary to induce a high population of the upper state at the expense of the lower (a population inversion). This is effected in a optical cavity by enclosing the lasing material between a pair of parallel mirrors and applying excitation either in the form of a flash of white light or an electric discharge. Light emitted during the spontaneous process is reflected between the plates inducing the stimulated emission which brings about further emission in a kind of chain process. Population of the upper state achieved using a flash lamp is called optical pumping, and the relevant energy levels of a typical solid-state laser (ruby) and a schematic drawing of the laser are given in Figs. 5.5 and 5.6 respectively.

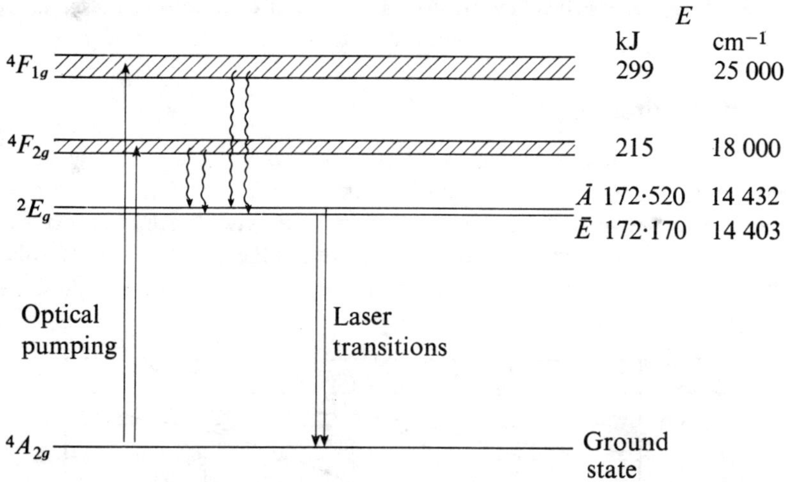


Fig. 5.5 Energy levels in the ruby (0·05% $Cr^{3+}$) laser.

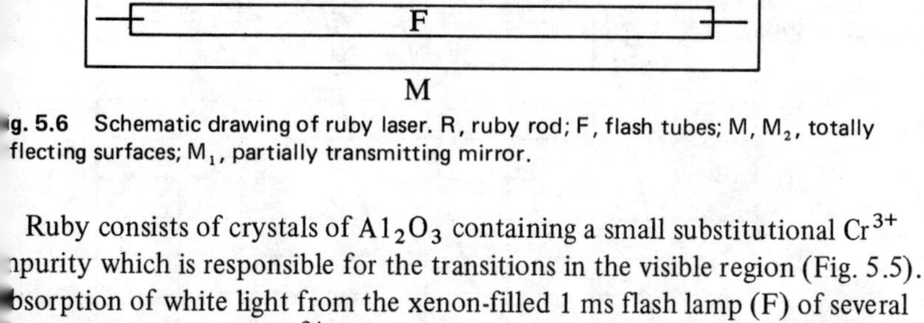


ig. 5.6 Schematic drawing of ruby laser. R, ruby rod; F, flash tubes; M, $M_2$, totally flecting surfaces; $M_1$, partially transmitting mirror.

Ruby consists of crystals of $Al_2O_3$ containing a small substitutional $Cr^{3+}$ npurity which is responsible for the transitions in the visible region (Fig. 5.5). bsorption of white light from the xenon-filled 1 ms flash lamp (F) of several energy excites many $Cr^{3+}$ atoms in the ruby rod (R) to the $^4T_{1g}$ and $^4T_{2g}$ ates. Intersystem crossing to the narrowly separated $^2E_g$ states is followed by oth spontaneous and stimulated return to the $^4A_{2g}$ ground state, mostly from e lower $\bar{E}$ state at 14 403 $cm^{-1}$ (694·3 nm), with a pulse duration of *ca.* 1 ms.

The light is subjected to multiple reflections in the optical cavity, which is elliptical in cross-section in one design, before being transmitted by a partially reflecting mirror ($M_1$). $M_2$ is a totally reflecting mirror.

Alternative laser systems involve a more complex distribution of energy levels, in which the laser transition is not to the ground state, or the use of gases which are 'pumped' by means of an electric discharge operating at radio frequencies. Four-level lasers, involving the interposition of an extra electronic level above the ground state, offer several advantages over the three-level ruby laser including higher efficiency. Whilst the ruby laser operates in a pulsed mode, the gas lasers can be operated continuously but at much reduced power.

## 5.2 Optical filters

Isolation of a particular wavelength is often as necessary in preparative work, where it is desired to prevent the product undergoing further transformations, as it is in quantitative work in which, say, the primary photochemical process is being examined. This goal may be achieved either by the use of a commercially available monochromator or by use of an interference or transmission filter.

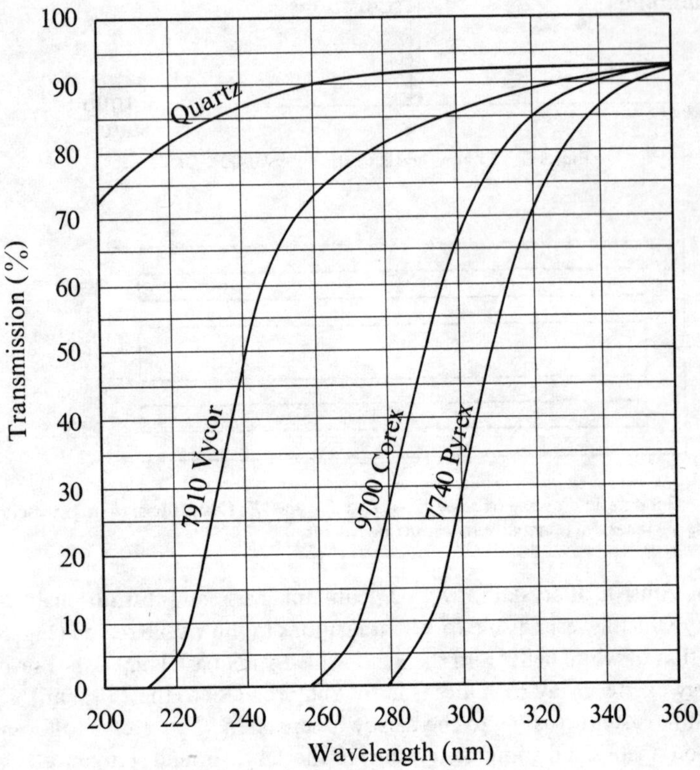


**Fig. 5.7** Transmission curves for various glasses (2 mm layer). (Reproduced by permission of Englehard Hanovia Inc.)

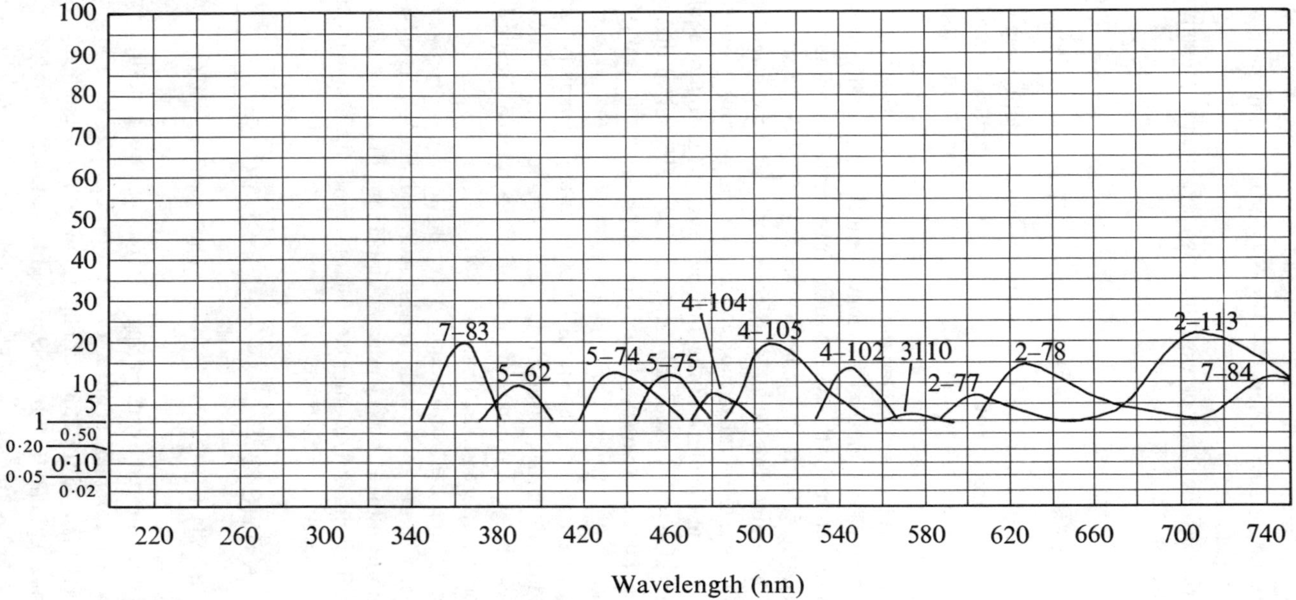


**Fig. 5.8** Transmittance curves of narrow band pass filter. (Reproduced by permission of Corning Glass International.)

Transmission filters restrict the full arc spectrum to certain wavelengths characteristic of the absorption properties of the various glasses. This absorption may be total below certain wavelengths (Fig. 5.7), or may allow passage of narrow bands (Fig. 5.8).

A whole variety of chemical solution filters is available, an excellent summary of which has been made by Kasha.[6]

## 5.3 Actinometry

Determination of the quantum yield for a photochemical reaction depends both on an analytical determination of reactant consumed or product formed during a given period of irradiation and on an accurate measurement of the total light absorbed by the system at the wavelength utilized during this period. Provided either *all* the light incident upon the sample is absorbed or a correction for transmitted light is made, then the most convenient and universally adopted procedure is to substitute for the system under investigation a suitable sample of material of clearly defined photochemical behaviour and to measure its response in a comparable time under identical experimental conditions. Such a material is known as a chemical actinometer and several systems have been utilized, depending on the phase of the unknown system (gaseous or liquid) and the wavelength region of the exciting light. Care must be exercised that the 'unknown' system does not absorb all incident light in a tiny element of solution close to the window of entry.

Desirable characteristics of a chemical actinometer are availability, thermal stability, reproducibility, high absorption, and uniform response over a relatively wide range of wavelengths, and ease of monitoring the induced chemical change, which should be 'clean', i.e., free of side-reactions.

For solution work in the 'conventional' wavelength range 250–435 nm, the uranyl oxalate and, more recently, the ferrioxalate actinometers have been widely used, although the latter is gradually supplanting the former, being much more sensitive and offering an extended range into the visible region up to 578 nm. The relevant reactions are

$$K_3Fe(C_2O_4)_3 \xrightarrow{h\nu} Fe(II) + CO_2 \quad (5.1)$$

$$U(VI) + (CO_2H)_2 \xrightarrow{h\nu} U(VI) + CO_2 + CO \quad (5.2)$$

Reaction (5.1) is monitored by measuring Fe(II) absorptiometrically as its 1,10-phenanthroline complex, addition of the complexing agent following exposure. $\varphi$(Fe(II)) is 1·25 for 253·7 nm light, but decreases gradually to about unity for 436 nm light and 0·86 for 509 nm light, being also slightly sensitive to ferrioxalate concentration. Full details of the preparation of the solution and the quantum yields are given by the inventors, Hatchard and Parker,[7] and are summarized by Calvert and Pitts.[8]

Reaction (5.2) is normally followed by titrating unreacted oxalate with permanganate, a necessarily rather insensitive procedure. $\varphi$(-oxalate) is in the region of 0·58 at wavelengths 245, 265, 278, 302, 313, 405, and 435·8 nm but departs a little from this figure at certain other wavelengths. Details of solution preparation and handling and quantum yields are provided by Masson, Bockelheide, and Noyes.[9]

With development of interest in the photochemistry of transition-metal complexes, there has arisen a need for a reliable actinometer operating at wavelengths above 600 nm and Adamson[10] has suggested that this need might be fulfilled by Reineckate ion, which absorbs throughout the visible to react as follows

$$[trans\text{-}Cr(NH_3)_2(NCS)_4]^- \xrightarrow{h\nu} NCS^-$$

with a quantum yield of 0·31 (452 nm), 0·29 (520 nm), and 0·27 (750 nm). $NCS^-$ is estimated absorptiometrically as its Fe(III) complex. $Cr(urea)_6^{3+}$ is also promising as a long-wavelength actinometer,[10] undergoing photoaquation with a quantum yield of *ca.* 0·095 at 652, 676, 696, and 735 nm.

The variety of molecules which have been used in gas-phase actinometry is illustrated in Table 5.1.

TABLE 5.1: Gas phase chemical actinometric substances

| Actinometer molecule | $\lambda$ (nm) | Product analysed | $\varphi$ | Conditions |
|---|---|---|---|---|
| Acetone | 250–320 | CO | 1·0 | $T > 400$ K, $p < 50$ mm |
| Diethyl ketone | 250–320 | CO | 1·0 | $T > 400$ K, $p < 50$ mm |
| HBr | 180–250 | $H_2$ | 1·0 | $T \sim 300$ K, $p \sim 10^2$ mm |
| $O_2$ | 130–190 | $O_3$ | 2·0 | $p \sim 1$ bar. Fast flow system necessary |
| $N_2O$ | 147–184·9 | $N_2$ | 1·44 | $T \sim 300$ K, $p$ 1–35 mm |
| $CO_2$ | 170 | CO | 1·0 | Flow system |

It should be appreciated that chemical actinometers, whilst accurate and convenient, are essentially secondary standards. The primary standard of radiant energy measurement is a thermopile coupled with a galvanometer, which is calibrated with a standard light source.

## 5.4 Conventional photolysis procedures

The experimentation required is determined by the objective. For preparative work the aim is to expose the sample to the maximum quantity of light in order to secure an effective degree of conversion in the shortest possible time. This aim is met with in the simple photochemical reactor, which comprises a low- or medium-pressure mercury arc cooled in a quartz or pyrex water-jacket and immersed in the reactant (Fig. 5.9).

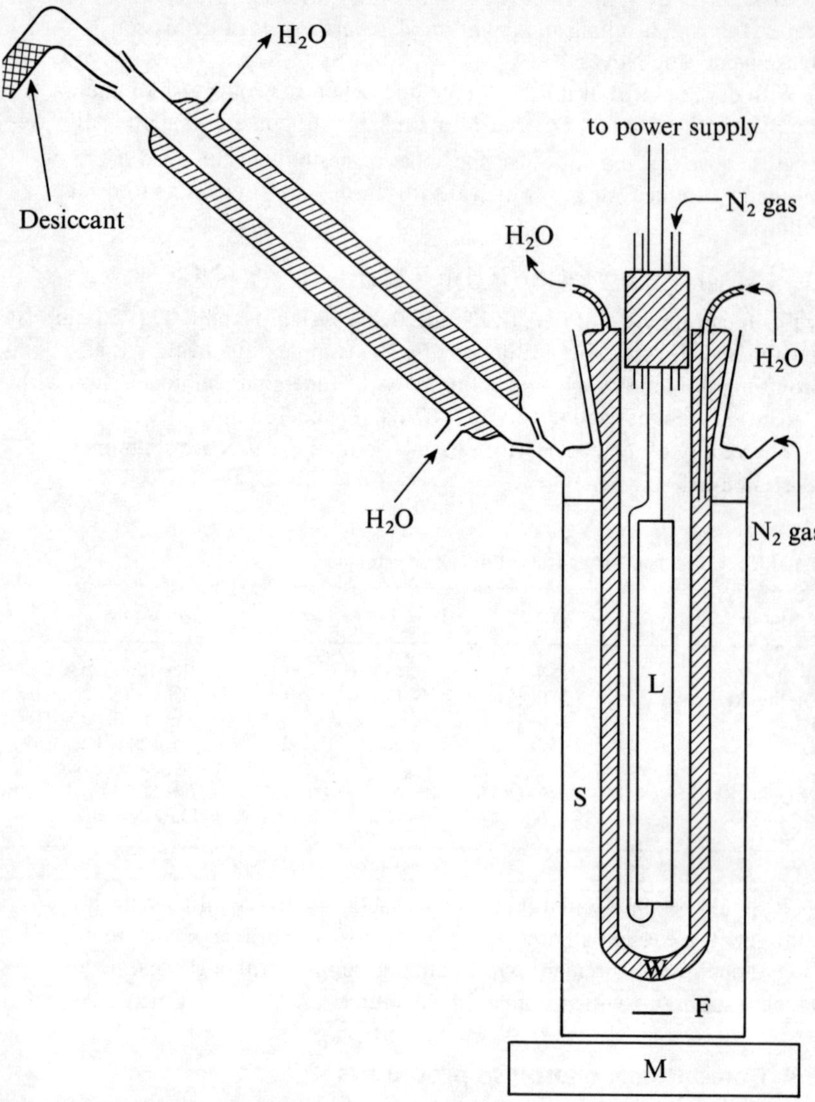


**Fig. 5.9** Photochemical reactor; L = lamp, W = water, S = solution under irradiation, F = follower, M = magnetic stirrer.

For quantitative work, particularly quantum yield determination, more detailed attention to geometry must be paid (Fig. 5.10). It is necessary to employ

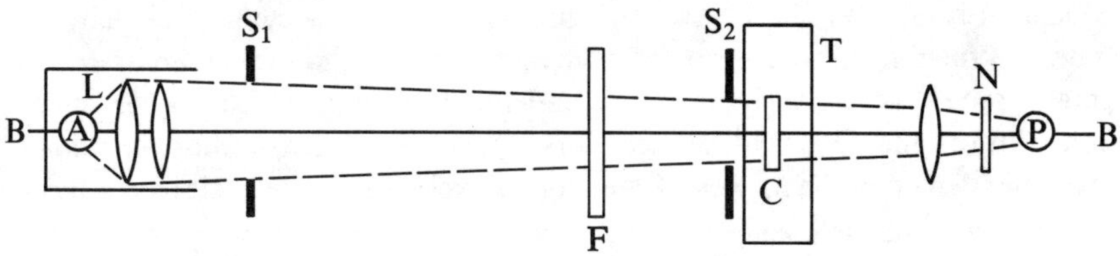


**Fig. 5.10** Photochemical bench (B) incorporating A, source; L, lenses $S_1, S_2$, light stops; F, filter; C, sample cell; T, thermostat; N, neutral density filter; P, photodetector.

an optical bench (B) on which is mounted the sequence; source (A), lens system (L) light shutter with circular opening ($S_1$), filter (F), shutter ($S_2$), sample cell (C) in a thermostat (T) with quartz windows. After passage through the sample the light is attenuated by means of a neutral density filter (N) before being directed onto a photocell (P), which monitors any fluctuation in the exciting light. Alternatively a quartz plate can be inserted at an angle to the light beam before it reaches the cell, reflecting a small component to the monitoring system. Materials are normally of fused quartz unless interest is confined to the visible and near u.v., when pyrex will suffice, or when work below 200 nm is to be undertaken, necessitating the use of LiF or $CaF_2$ windows. The cell is normally cylindrical, typically of 20 mm pathlength and 70 mm diameter with optically flat windows and with a side-arm leading to a bulb suitable for degassing the sample by freeze-pump-thaw cycles on a vacuum line. Other attachments to the cell may include a smaller quartz cell for allowing spectrophotometric examination of the sample from time to time.

The source is usually a mercury arc run off a current-stabilized power supply and the exciting line required is selected by means of the appropriate filter solution or glass (F) (section 5.2) or a monochromator.

Following examination of the sample, a chemical actinometer is placed in the cell (section 5.3).

## 5.5 Flash photolysis

This technique, developed by Norrish and Porter[11, 12] and recognized in the award to these workers of the 1967 Nobel Prize for Chemistry jointly with Eigen, has made such an impact in the elucidation of primary photochemical processes of both simple and complex molecules in solution and in the gas phase that no compilation of its applications is necessary and for an appreciation of its scope the reader is referred to chapters 2 and 3.

A typical experimental lay-out is shown in Fig. 5.11 and the principles of the method are as follows.

A flash of white light of duration *ca.* $10^{-6}$ s is produced by discharging a condenser bank (C) of energy *ca.* 1 kJ through one or more cylindrical quartz tubes (F) filled to a low pressure with an inert gas. Close to these tubes is a parallel tube (S) filled with the sample under investigation, which may be gaseous or a solution, and which will suddenly undergo a high degree of photoexcitation, markedly depleting the $S_0$ level. Detection of excited states or radicals produced in the sample tube is usually accomplished optically in one of two ways; (*a*) the spectrographic mode whereby a second, similar but weaker 'analysing' or 'spectrographic' flash (A) is triggered at a pre-set time after the primary or main flash by means of a delay unit (D) and traverses the length of the sample tube before being focused onto the entrance slit of a spectrograph (G); (*b*) the kinetic-spectroscopic or photoelectric mode in which the spectrographic flash is replaced by a continuous light source, and the spectrograph by a monochromator–

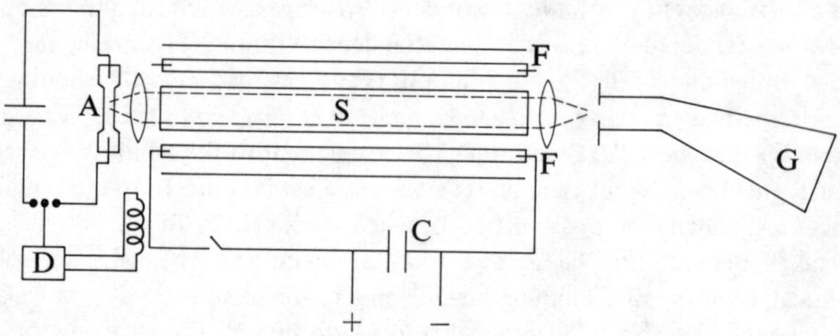


**Fig. 5.11** μs Flash photolysis apparatus. S, sample cell; F, F, flash tubes; C, condenser bank; D, delay unit; A, spectrographic flash tube; G, spectrograph.

photomultiplier tube–cathode-ray oscilloscope system. An absorbing transient species in S temporarily reduces light received by the photomultiplier tube and the recovery of the current provides a register of the kinetics of the decay of the transient absorption.

Normally the spectrographic mode is employed in a preliminary investigation, in order to establish the absorption spectrum of any accessible intermediates, and the kinetic-spectroscopic mode is then used to obtain kinetics of decay of these intermediates at their wavelength maxima and the kinetics of growth of any new species arising in the system.

By taking a number of spectrographic recordings at a series of different time delays between firing the primary and spectrographic flash tubes it is possible to obtain both spectroscopic and kinetic records of the progress of the early stages of the photochemical reaction.

In a recent development, Porter and others[13,14] have used a Q-switched laser source to obtain flashes of *ca.* $10^{-9}$ s duration. For most molecules of interest ultra-violet laser light is necessary and suitable lasing materials include nitrogen (λ 337·1 nm), ruby (λ 694 nm), and neodymium (λ 1060 nm). The outputs from the last two are frequency doubled and quadrupled respectively, albeit with a considerable loss of energy, by passage through orientated crystals of ammonium or potassium dihydrogen phosphate. On such short time-scales major problems of detection have to be overcome including the provision of sufficient light intensity in a continuous monitoring beam and the devising of a suitably short analysing flash, both covering a wide wavelength range and capable of being fired at a few ns after the exciting flash. One ingenious approach[14] to these problems, depicted in Fig. 5.12, involves the use of a beam-splitter (B) which

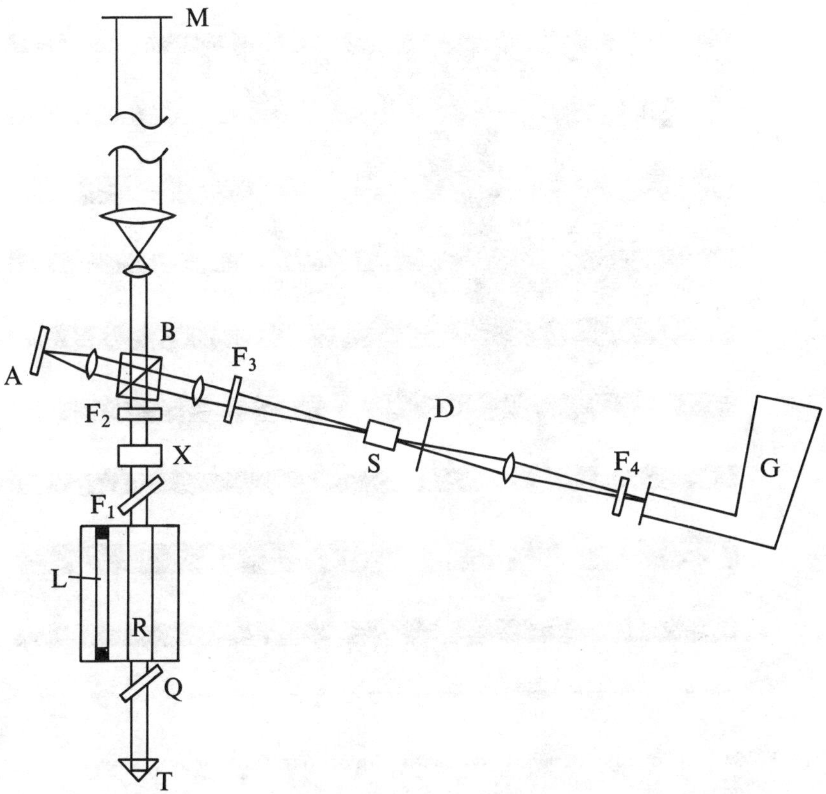


**Fig. 5.12** Nanosecond laser flash spectrographic apparatus of Porter and Topp.[14] A, scintillator cell; B, beam splitter; D, aperture stop; F, filters; $F_1$, Wratten 29, transmits $\lambda \geqslant 630$ nm; $F_2$, $CuSO_4$, attenuates 694 nm; $F_3$, biphenylene, control of intensity of 347 nm; $F_4$, biphenylene, u.v. laser cut-off; G, spectrograph; L, xenon-filled flash lamp; M, mirror; Q, passive Q-switch; R, ruby laser; S, reaction vessel; T, quartz t.i.r. prism; X, ADPh crystal. (Reproduced with permission of the authors and publisher.[14])

directs a portion of the laser light into the sample tube (S) and the remainder to a mirror (M) fixed at a pre-set distance. The returning component re-enters the beam splitter and is reflected to a scintillator solution (A) which fluoresces over a few ns at longer wavelengths, providing a suitable flash which is directed through the reaction vessel (S) to the spectrograph (G). For photoelectric work, continuous monitoring of the reaction vessel cannot be performed with a conventional lamp because of diminished signal-to-noise ratios on the time-scale involved, and a pulsed source is required.

Whilst μs flash photolysis has enabled the accumulation of an enormous body of information on free radicals in the gas phase and solution and on triplet

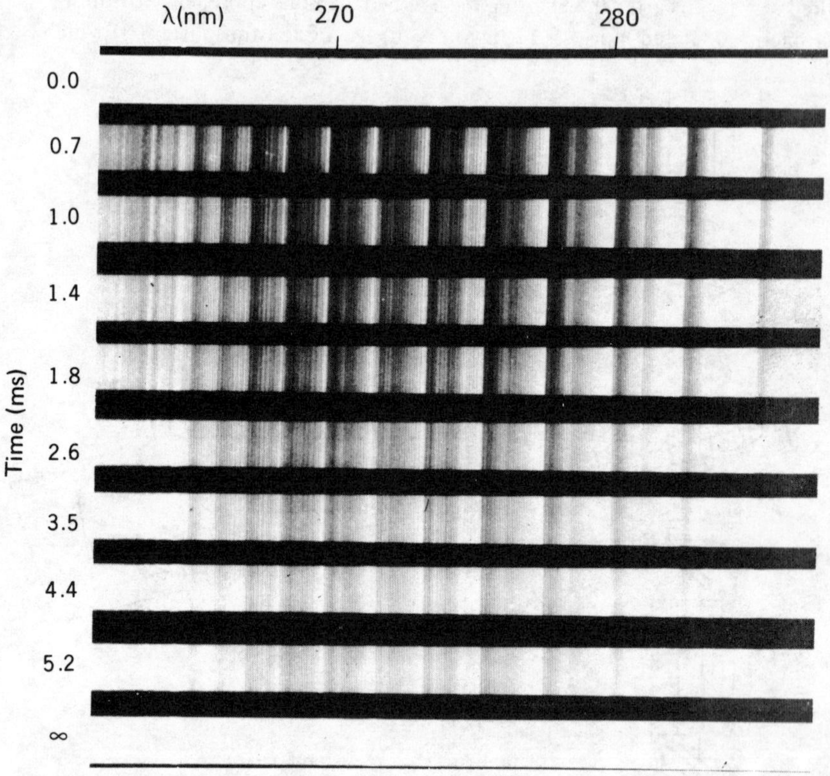


**Plate I** ClO˙ bimolecular decay. (Reproduced with permission of the authors and publisher.[15])

excited states, especially as regards their absorption spectra, reactions, lifetimes, and quenching processes, the development of ns systems has extended the field of study to singlet states (normally $S_1$) along parallel lines. The diverse achievements of both μs and ns systems are illustrated by Plates I–IV. Plate I shows a series of spectrographic recordings of the ClO˙ radical following flash photolysis of a $Cl_2 + O_2 + N_2$ mixture[15] which exhibit the second-order decay of this species over a period of ms. In Plate II are shown the absorptions of a $10^{-5}$ M

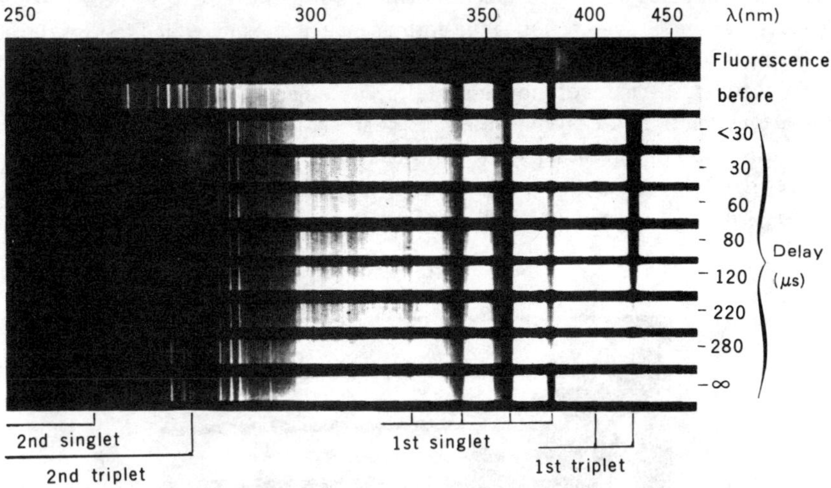


**Plate II** Triplet spectra of anthracene in hexane. (Reproduced with permission of the authors and publisher.[16])

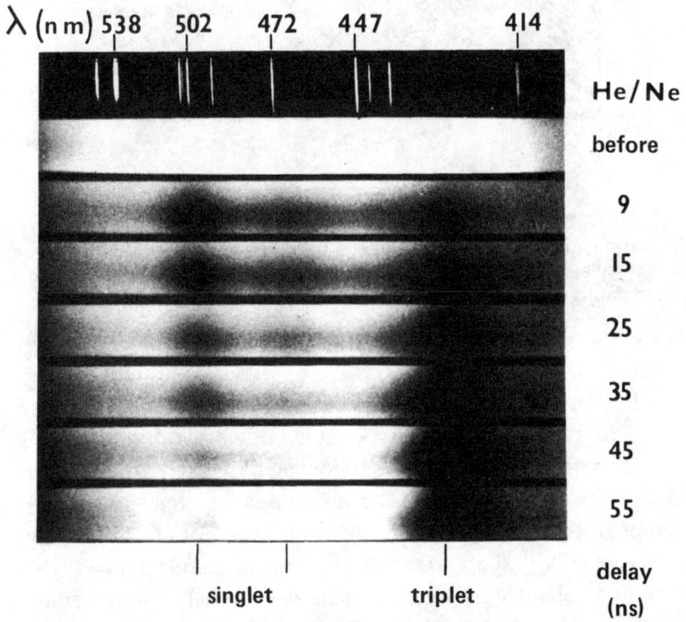


**Plate III** Absorptions of excited singlet and triplet triphenylene following laser flash photolysis in benzene. (Reproduced with permission of the authors and publisher.[14])

solution of anthracene in hexane following flash photolysis;[16] the decaying species in the 425 nm region is the anthracene triplet state which returns to the ground state to enhance the temporarily depleted $S_1 \leftarrow S_0$ absorption at 380 nm. Point-to-point spectra of inorganic radicals obtained by μs flash photolysis of aqueous solutions are given in Figs. 2.8(a) and (b).

Absorption from the $S_1$ state of triphenylene in benzene obtained in ns laser flash photolysis is shown in Plate III, the decay of the $S_2 \leftarrow S_1$ *absorption* of $t_{1/2}$ 44·2 ns agreeing well with the fluorescence decay time of 43·0 ns.[14] Plate

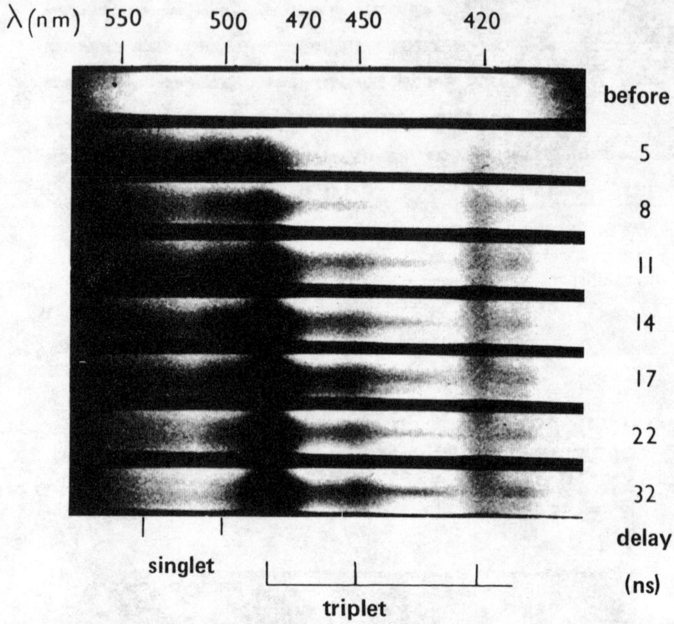


**Plate IV** Absorptions of excited singlet and triplet phenanthrene following laser flash photolysis of an aerated solution in cyclohexane. (Reproduced with permission of the authors and publisher.[14])

IV shows the oxygen-perturbed conversion of $S_1$ to $T_1$ states for a solution of phenanthrene in cyclohexane.[14]

Whilst the foregoing discussion has dealt with *absorption* of excited states following flash photolysis, there is a related body of work concerned with *emission*. Detection of fluorescence following flash excitation is much easier than that of absorption as no monitoring light is required and a weaker flash lamp can be utilized, and an impressive compilation of lifetimes of $S_1$ states, which are generally in the $1\text{–}10^2$ ns region, has accumulated since 1955. Also related and complementary to these techniques is that of pulse radiolysis in which a μs or ns pulsed beam of ionizing radiation replaces the flash lamp.

An important extension of the emission work to the ps region is due to Rentzepis.[17] ps Pulses from a Q-switched neodymium-glass laser (λ 1060 nm) were passed through a potassium dihydrogen phosphate crystal to give some

530 nm photons. Because of the differing group velocities in the 530 and 1060 nm pulses after passage through a dispersing cell, the latter enters the sample cell, containing a solution of azulene, first, but is not absorbed and is reflected from a mirror placed at the rear of the cell to encounter the slower 530 nm pulse. At the point of collision intense $S_2 \rightarrow S_0$ fluorescence of azulene is detected which is not produced by the 530 nm pulse alone. The combined photon energies are equivalent to 355 nm, coinciding with the $S_2$ level of azulene and clearly two-photon absorption is taking place, with the 530 nm pulse populating the non-fluorescent $S_1$ level in vibrationally excited states and the returning 1060 nm pulse promoting $S_1$ state molecules to the fluorescent $S_2$ level. By delaying return of the 1060 nm photon, relaxation in the $S_1$ state to lower vibrational levels with insufficient energy to be excited to the $S_2$ level occurs and from the resulting emission intensities a value for this relaxation time can be obtained, e.g., from the $S_1$ $(v = 4)$ level to the $S_1$ $(v = 2)$ level, $\tau = 7{\cdot}5 \times 10^{-12}$ s.

## 5.6 Fluorimetry and phosphorimetry

Development of the techniques of measurement of fluorescence and phosphorescence spectra has received considerable impetus from the analytical possibilities, particularly of fluorimetry. Fluorescence spectra are measured with a spectrofluorimeter (Fig. 5.13) which consists of an excitation source (A) providing a

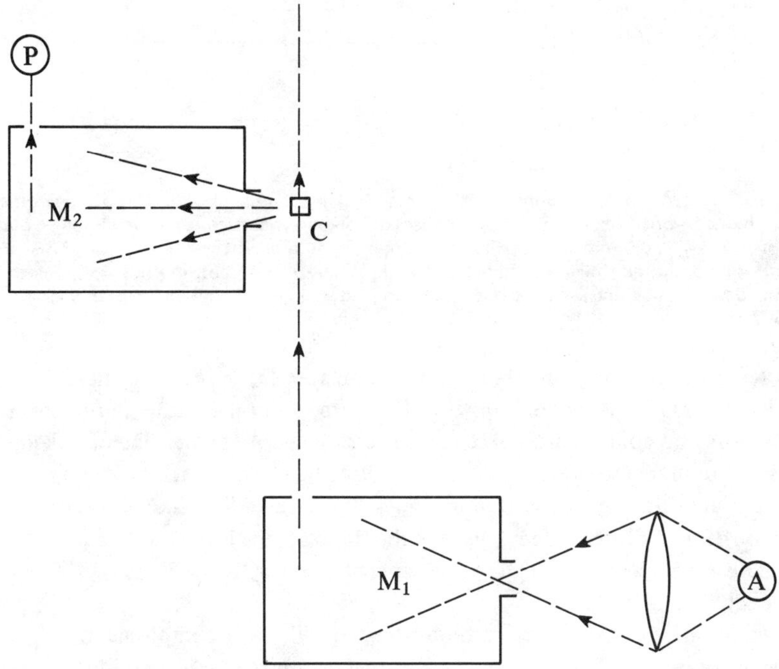


Fig. 5.13 Layout of spectrofluorimeter. A, excitation source; $M_1$, $M_2$, monochromators, C, cell; P, photomultiplier tube.

continuum of light focused onto the slits of a monochromator ($M_1$) from which light is reflected to the sample cell (C).

Sample fluorescence passes into a second monochromator ($M_2$) and thence to a photomultiplier tube (P). Accordingly variation of $M_2$ provides the spectral distribution of fluorescence emission whilst the change of fluorescence intensity (at the wavelength maximum of the fluorescence spectrum) with wavelength of the exciting light–the fluorescence excitation spectrum–is obtained by variation of $M_1$. The use of two monochromators implies that only a tiny fraction of the output of A reaches P and one must use either wide slit settings on $M_1$ (for

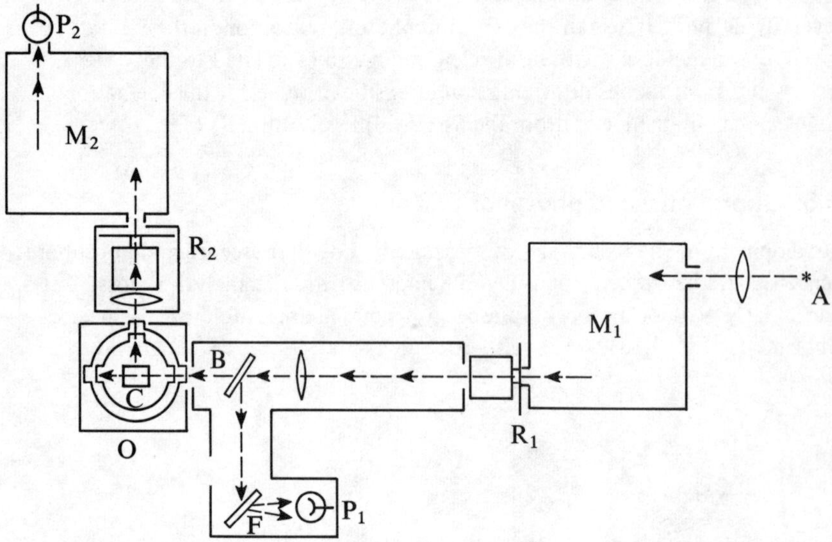


**Fig. 5.14** Layout of spectrophosphorimeter.[19] A, light source; $M_1$, $M_2$, Hilger D247 quartz prism monochromators; $R_1$, $R_2$, chopper-discs driven by synchronous motors; B, silica plate beam splitter; F, 0·5 mm silica optical cell containing fluorescent screen solution; $P_1$, monitoring photomultiplier; $P_2$, fluorescence-phosphorescence photomultiplier; Q, fused quartz dewar containing sample cell C. (Reproduced with permission of the authors and publisher.[19])

fluorescence spectra) or on $M_2$ (for excitation spectra) or resort to the use of filters instead of one monochromator. For purposes of measuring fluorescence spectra only, a continuous source is unnecessary and A can be a mercury lamp. It is also desirable to prevent scattered exciting light from reaching P and an appropriate filter can be inserted in front of $M_2$. An additional complication is due to the 'inner filter' effect whereby the fluorescence is re-absorbed by ground-state molecules at high solute concentrations. Other effects are discussed in detail by Parker.[18]

Quantitative measurement of fluorescence yields are accomplished by comparison with a fluorescence standard such as aqueous acidic quinine bisulphate ($\varphi_F = 0{\cdot}55$). Correction for wavelength-dependent factors such as photomultiplier tube response is effected by measuring the response to a source offering a fairly constant output over a wide spectral region.

Phosphorescence spectra can also be obtained using a double monochromator system, for example that of Parker and Hatchard,[19] (Fig. 5.14). The emission is much longer-lived than fluorescence and the two are separated by viewing the sample in a period after expiry of the latter. The requisite delay between excitation and measurement is achieved mechanically by chopping both the light from the source (A) and the emission from the refrigerated sample (C); the two choppers ($R_1$,$R_2$), consisting of discs with a series of holes or slots cut out of them, are operated out-of-phase with each other by synchronous motors so that emission is passed by $R_2$ to the detector ($P_2$) whilst the source is shielded by $R_1$ during one of its dark periods. By operating the choppers in-phase the sum of the fluorescence and phosphorescence is obtained. In simpler instruments the sample is positioned inside a rotating can having a single vertical slot cut in its side. During one rotation the slot firstly admits light from the source and then passes emission to the analysing monochromator. $\varphi_P$ is determined in much the same manner as $\varphi_F$, using a fluorimetric standard.

## References


1. G. Ciamician and P. Silber, *Ber. dtsch. chem. Ges.*, **41**, 1928 (1908).
2. H. E. Gunning and O. P. Strausz, *Adv. Photochem.*, **1**, 209 (1963).
3. P. Ausloos, National Bureau of Standards Technical Note 496, *Rare Gas Resonance Lamps*, U.S. Department of Commerce, Washington, D.C. (1969).
4. A. Schönberg, G. O. Schenck, and O. A. Neumüller, *Preparative Organic Photochemistry*, Springer-Verlag, Berlin, p. 472 (1968).
5. T. H. Maiman, R. H. Hoskins, I. J. D'Haenens, C. K. Asawa, and V. Evtuhov, *Phys. Rev.*, **123**, 1145, 1151 (1961)
6. M. Kasha, *J. Opt. Soc., Amer.*, **38**, 929 (1948).
7. C. G. Hatchard and C. A. Parker, *Proc. Roy. Soc.*, **A235**, 518 (1956).
8. J. G. Calvert and J. N. Pitts, Jr., *Photochemistry*, Wiley, New York, p. 783 (1966).
9. C. R. Masson, V. Boekelheide, and W. A. Noyes, Jr., 'Photochemical Reactions' in *Technique of Organic Chemistry*, Vol. II (2nd ed.), ed. A. Weissberger, Interscience, New York, p. 294 (1956).
10. E. E. Wegner and A. W. Adamson, *J. Amer. Chem. Soc.*, **88**, 394 (1966).
11. R. G. W. Norrish and G. Porter, *Nature* (Lond.), **164**, 658 (1949).
12. G. Porter, *Proc. Roy. Soc.*, **A200**, 284 (1950).
13. J. R. Novack and M. W. Windsor, *Proc. Roy. Soc.*, **A308**, 95 (1968).
14. G. Porter and M. R. Topp, *Proc. Roy. Soc.*, **A315**, 163 (1970).
15. G. Porter and F. J. Wright, *Disc. Faraday Soc.*, **14**, 23 (1953).
16. G. Porter and M. W. Windsor, *Disc. Faraday Soc.*, **17**, 178 (1954).
17. P. M. Rentzepis, *Chem. Phys. Lett.*, **2**, 117 (1968).
18. C. A. Parker, *Photoluminescence of Solutions*, Elsevier, Amsterdam, Chapter 3 (1968).
19. C. A. Parker and C. G. Harchard, *Trans. Faraday Soc.*, **57**, 1894 (1961).

# 6. Topics related to photochemistry

## 6.1 Photosynthesis[1]

### 6.1.1 The basic processes

On a world-wide scale it is estimated that $2 \times 10^{14}$ kg (approximately two hundred thousand million tons) of carbon are fixed annually in photosynthetic processes. This is accomplished by two groups of organisms, the green plants which depend upon water as photoreductant and the photosynthetic bacteria which require a more powerful reductant. In green plants and algae the three principal stages in photosynthesis are as follows.

(a) Absorption of energy by chlorophyll leading to a separation of oxidizing and reducing entities.
(b) Electron transfer reactions using these oxidizing and reducing entities as starting points and culminating in the storage of chemical energy as ATP, the storage of reducing power as reduced pyridine nucleotide and the release of oxygen from water.
(c) Utilization of the stored chemical energy and reducing power by conversion of $CO_2$ to essential cell constituents (e.g., sugar).

The overall photochemical reaction is

$$CO_2 + 2H_2{}^{18}O \xrightarrow{h\nu} [CH_2O] + {}^{18}O_2 + H_2O$$

Photosynthetic bacteria oxidize such substrates as $H_2$, $H_2S$, thiosulphate, and many organic compounds, and store chemical energy as ATP.

Van Niel[2] has been responsible for laying the foundations of an understanding of the primary events of photosynthesis and Fig. 6.1 summarizes his formulation.

Recent suggestions present a more sophisticated picture but the essence of van Niel's ideas is still accepted. In particular, Emerson[3] has shown that the quantum yield of photosynthesis, which is reasonably constant between 500 and 680 nm, drops severely at longer wavelengths (the Emerson 'red drop' phenomen[on] and that this low efficiency can be improved by simultaneously supplying light of wavelength $< 680$ nm. The phenomenon has been interpreted as meaning that wo photosystems, coupled by a series of electron carriers, are present, one uantum being necessary to activate each of them. Photosystem I (PS I), nergized by far red light, oxidizes the carriers and photosystem II (PS II),

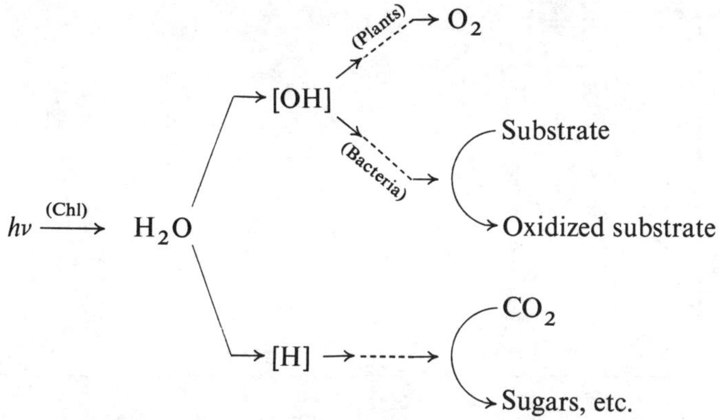


**Fig. 6.1** A representation of van Niel's formulation of photosynthesis.

ependent upon shorter wavelengths, reduces them. Chlorophyll a is believed be the major light harvesting pigment in PS II and chlorophyll b functions

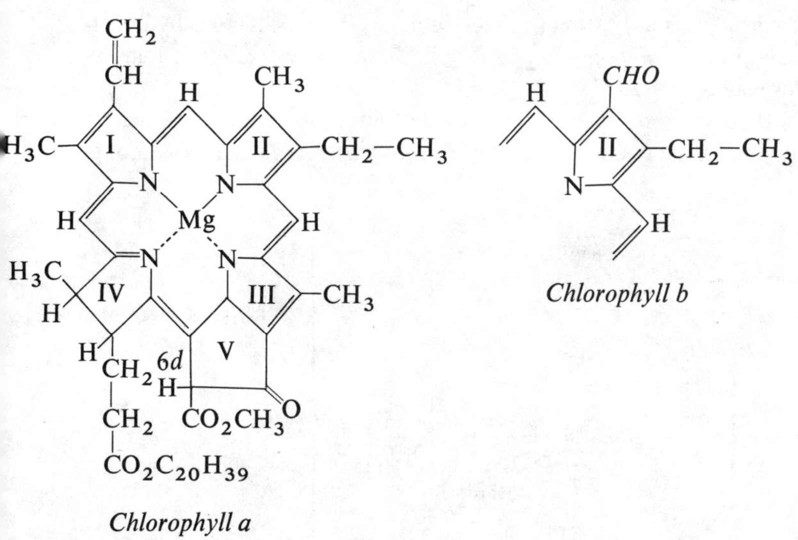


**Fig. 6.2** The structure of chlorophylls a and b.

milarly in PS I. The structures of chlorophylls a and b and their absorption ectra are shown in Fig. 6.2 and Fig. 6.3, respectively.

Precisely how these two systems co-operate in the cleavage of an H–O bond unknown, but there appears to be general agreement on major points as mmarized in Fig. 6.4.

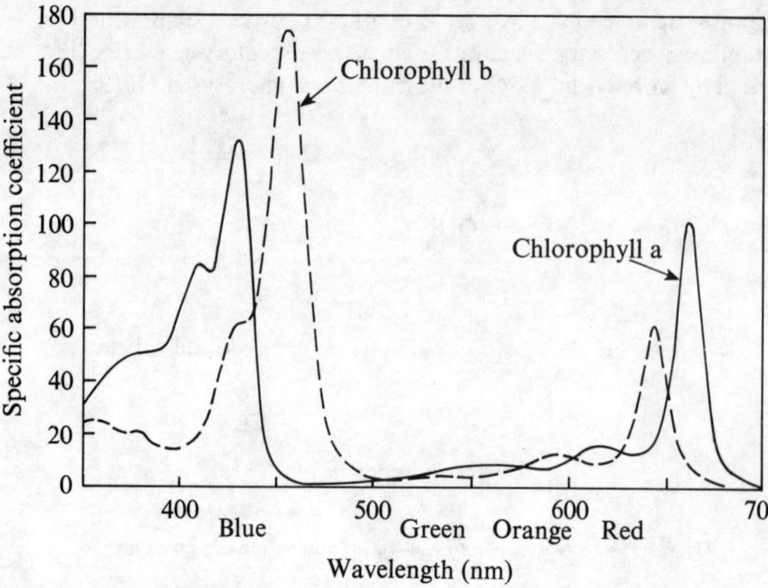


**Fig. 6.3** The absorption spectra of chlorophylls a and b in ether.

The mode of action of chlorophyll is somewhat obscure, but it is possible that following $\pi^* \leftarrow \pi$ excitation (cf. section 6.1.3) in PS II, there is hydrogen abstraction from carbon atom *6d* (Fig. 6.2). In this way electron transfer chains may be initiated.

PS II contains an electron acceptor Q of unknown structure, which quenches chlorophyll a fluorescence. As reduced Q, it transfers an electron through a

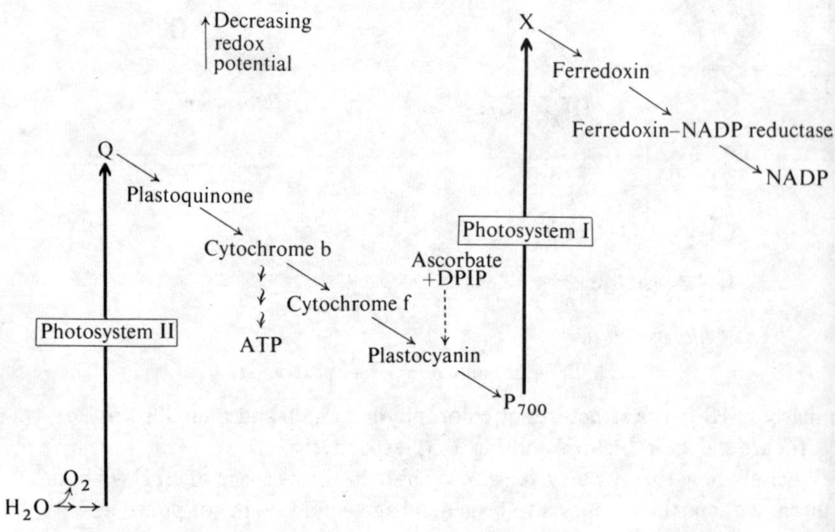


**Fig. 6.4** A scheme of photoinduced electron transport reactions in isolated chloroplasts.

series of components which include plastoquinone, cytochrome b, cytochrome f, and plastocyanin to $P_{700}$, a specific chlorophyll component which is the reaction centre of PS I.

Absorption, by PS I, of the second quantum results in electron transference to X, a low potential acceptor, again of unknown structure. This electron is transferred via a series of carriers, which includes ferredoxin and a flavoprotein, to one of the important cellular reductants, the pyridine nucleotides, $NADP^+$. This then participates in the photosynthetic carbon cycle, the details of which have been worked out by Calvin.[4] The overall reaction is

6 ribulose 1,5-diphosphate + $6CO_2$ + 18ATP + 12NADPH

$$\text{6 ribulose 1,5-diphosphate} + 6CO_2 + 18ATP + 12NADPH \rightarrow$$

Simultaneously, water is converted to oxygen in a transformation which, when it occurs in isolated chloroplasts, is called the Hill reaction.

### 6.1.2 Photosynthetic units

Chlorophyll molecules do not operate independently of each other *in vivo*, and the experiments of Emerson[5] have shown that, at least in the case of *Chlorella*, about 2400 molecules constitute a single photosynthetic unit. The absorption

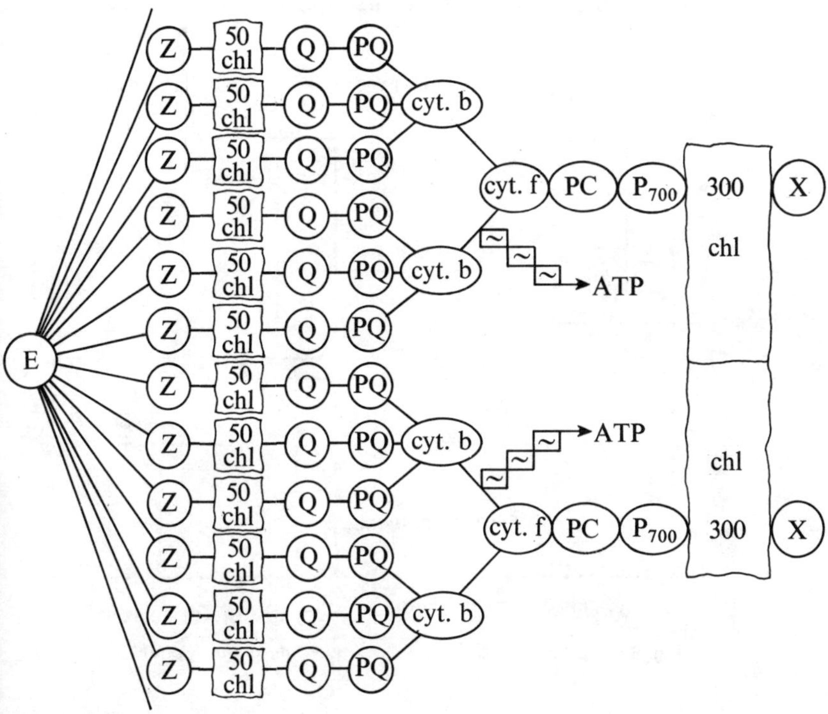


Fig. 6.5 A hypothetical model of the photosynthetic unit in isolated chloroplasts. Only half of a unit is illustrated. After Avron.[16]

of a minimum of eight quanta by this unit results in the reduction of one molecule of $CO_2$ and evolution of one molecule of oxygen. Thus 300 chlorophyll molecules act as a single reaction centre and are responsible for the transfer of one electron. An attempt has been made in Fig. 6.5 to represent a photosynthetic unit schematically.

All the 'Z' molecules are linked through a component 'E'[6] which constitutes the bottleneck of the unit.

*In vivo*, chlorophyll molecules occur in a 2- or 3-dimensional molecular aggregate, but only about 0·1% are photochemically active. The remaining 99·9% and the carotenoids harvest the incident radiation and deliver it to the photochemically active chlorophyll molecules. The mechanism by which this is accomplished is a consequence of the interactions present in molecular aggregate These interactions may result from intermolecular orbital overlap, or may be dipolar in nature and lead to electrical conductivity and exciton migration, respectively.

### *6.1.3 Excited states of chlorophyll*

While it is not possible to make detailed assignments it seems fairly well established that the major absorption bands at 430 and 660 nm in chlorophyll a

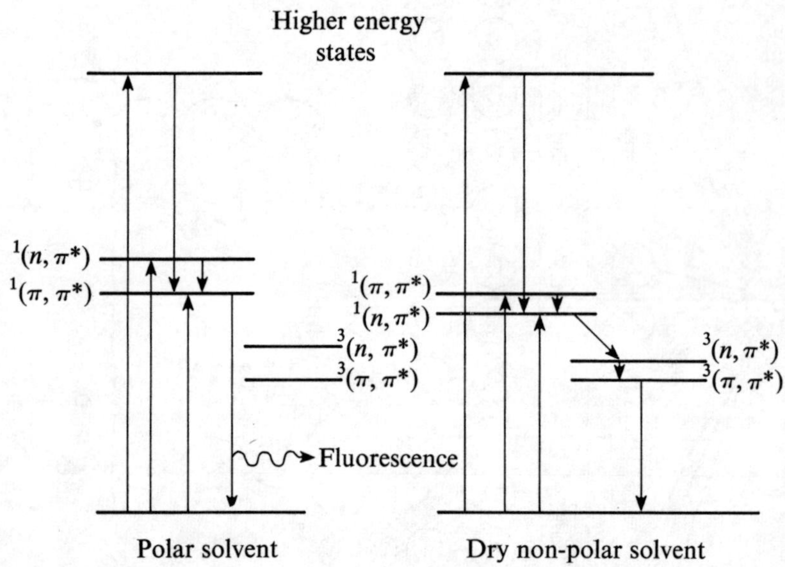


**Fig. 6.6** Solvent effects on the energy levels of chlorophyll.

result from $\pi^* \leftarrow \pi$ transitions to singlet states. Evidence for the existence of $(n, \pi^*)$ states is not so clear cut, however, although the shoulder at 670 nm next to the main band at 650 nm in the spectrum of a dry benzene solution of chlorophyll b could be due to such a state. Addition of water results in the disappearance of this shoulder, presumably due to a blue shift.

Chlorophyll b shows maxima in its phosphorescence spectrum at 733 and 865 nm probably corresponding to radiative transitions from the $^3(n, \pi^*)$ and $^3(\pi, \pi^*)$ states, respectively. Chlorophyll a displays a phosphorescence maximum at 755 nm ($^3(n, \pi^*)$) only, but Kasha[7] believes a $^3(\pi, \pi^*)$ state around 870 nm also exists.

Chlorophyll a solvates readily (probably at the central magnesium atom) and is only observed to fluoresce under these circumstances. This is rationalized on the basis of the relative energies of the $^1(n, \pi^*)$ and $^1(\pi, \pi^*)$ states. When the molecule is solvated, the $^1(\pi, \pi^*)$ level will be the singlet state of lowest energy and it will decay radiatively. In dry non-polar solvents, the $^1(n, \pi^*)$ state is lower than the $^1(\pi, \pi^*)$ state and intersystem crossing competes efficiently with fluorescence (Fig. 6.6).

Evidence that sensitized fluorescence is an important process in the functioning of a photosynthetic unit is good, in that, if photosynthesis is impeded, or if all the absorbed light cannot be used in photosynthesis, the fluorescence yield increases.

### 6.1.4 *The solution photochemistry of chlorophyll* in vitro

Chlorophyll can photosensitize a variety of reactions all of which probably involve triplet states. They can be classified as follows.

(a) Photooxidations of compounds such as alcohols, hydrocarbons, and chlorophyll itself.
(b) Photoreduction of chlorophyll (to yield a pink product, $\lambda_{max}$ 523 nm) by such compounds as vitamin C, $H_2S$, and phenylhydrazine.
(c) Two-stage electron transfer reactions involving vitamin C, $H_2S$, and phenylhydrazine as electron donors. In these reactions reduced chlorophyll is an intermediate and typical acceptors are azo dyes, *o*-dinitrobenzene, riboflavin, $NAD^+$, and $NADP^+$.
(d) Isomerization of poly-*cis*-carotenes.

Processes (b) and (c) may be important in photosynthesis.

Chlorophyll in dry non-polar solvents is relatively inert but is photochemically active in water. The excited state *in vitro* is almost certainly the $^3(\pi, \pi^*)$ state with a lifetime of the order of several milliseconds.

## 6.2 Chemiluminescence

### *6.2.1 Introduction*

Luminescence is any radiative process involving an electronically excited molecule. Chemiluminescence is a special case in which the electronically excited state has been attained in a chemical reaction and bioluminescence is the analogous process in living systems. It should be noted, however, that chemiluminescence and bioluminescence are often intended to imply radiative emission in the visible only, but this seems to be unnecessarily restrictive. Several excellent reviews of these topics are available.[8, 9, 10]

There is great structural diversity in the classes of compounds capable of exhibiting chemiluminescence and some typical examples are shown in Table 6.1.

TABLE 6.1: Examples of chemiluminescent reactions

(a)
$$\mathrm{Cl.CO.CO.Cl} \xrightarrow{H_2O_2} 2HCl + 2CO + O_2 + h\nu$$

(b)
$$\mathrm{R{-}MgBr} \xrightarrow{O_2} \text{Products} + h\nu$$

If R = *p*-bromophenyl, then *p*-bromophenol and bromobenzene are isolated as products. The mechanism possibly involves peroxide decomposition or electron transfer.

(c)

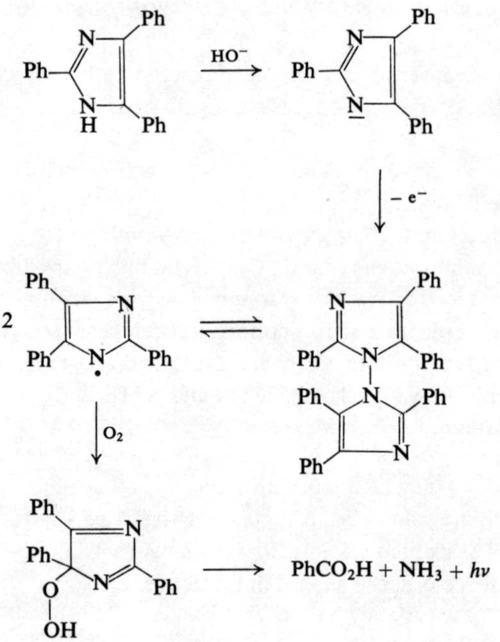


Isolated

(d)

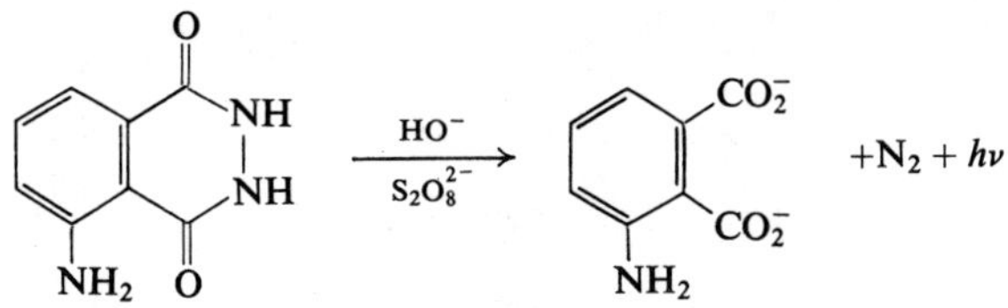


Luminol

(e)

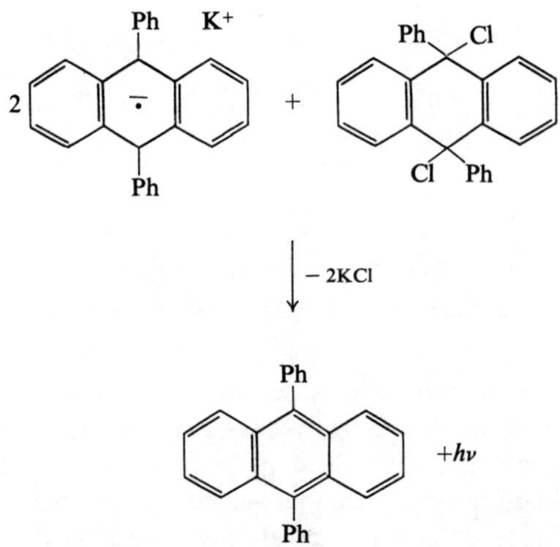


(f)

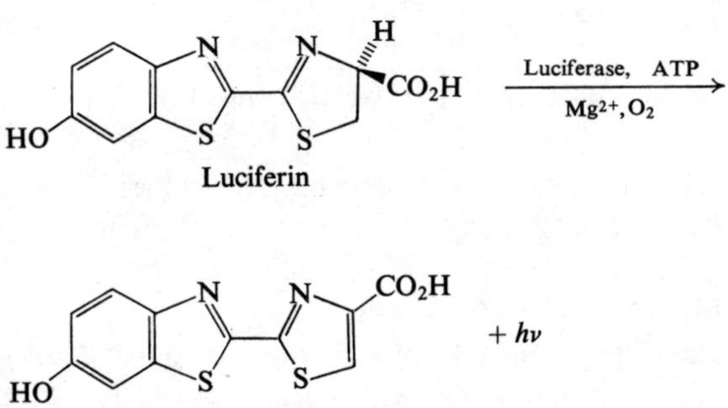


Luciferin is the active constituent of the American firefly (*Photinus Pyralis*).

Originally, it was believed that all chemiluminescent reactions require either molecular oxygen or a peroxide but this is now known to be untrue, and simple electron transfer processes can be effective in promoting chemiluminescenc (e.g., reaction (e)).

### 6.2.2 General mechanistic considerations

In general, a cheminluminescent reaction can be conveniently divided into three discrete steps.

(a) Formation of an intermediate.
(b) Conversion of the chemical energy of this intermediate into electronic energy.
(c) Radiation of this energy.

The efficiency of chemiluminescence, $\varphi_{CL}$, is consequently a product of two separate efficiencies, $\varphi_{ES}$ and $\varphi_L$,

$$\varphi_{CL} = \varphi_{FS} \cdot \varphi_F$$

where $\varphi_{ES}$ = Efficiency of production of the excited state
$\varphi_F$ = Efficiency of production of the fluorescence.

In this regard, it is convenient to consider the excited species as one of several possible products, each arising from the reactants by competitive pathways. Its yield is accordingly determined by the relative efficiencies of these pathways just as in any chemical reaction. Similarly, luminescence is in competition with processes such as internal conversion and external quenching. The quantum yield of the chemiluminescent reaction, $A + B \rightarrow C$, is given by,

$$\varphi_{CL} = \frac{\text{Photons emitted (einsteins)}}{\text{Mol of A (or B) consumed}}$$

For the emission of blue light ($\lambda \simeq 450$ nm), 265 kJ mol$^{-1}$ must be available and, correspondingly, 199 kJ mol$^{-1}$ for red light ($\lambda \simeq 600$ nm). However, an exact correlation between the energy of the radiation and the enthalpy of the reaction is not necessary as Fig. 6.7 shows.

Transfer of this energy must take place in one step and cause electronic excitation. Since chemiluminescence is a relatively rare phenomenon, it is apparent that energy transfers which do occur are normally to the vibrational levels of the product molecules. This difficulty can be overcome by utilization of a highly exothermic reaction, leading to a species with only limited vibrational

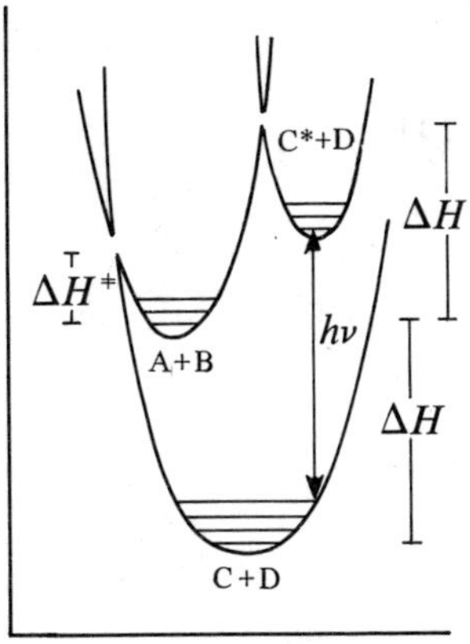


Non-chemiluminescent
(Product fluorescent)

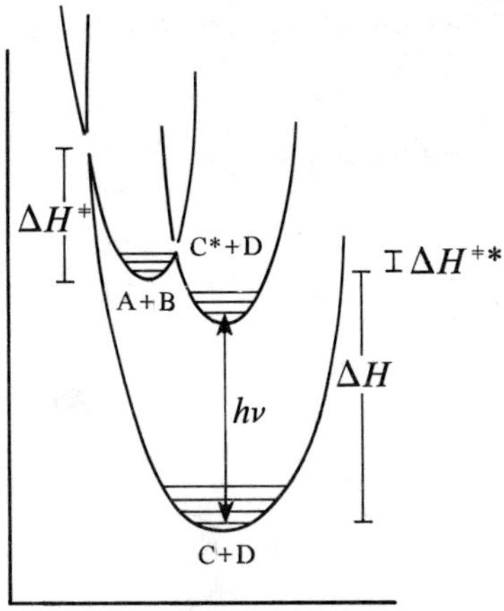


Chemiluminescent
reaction
$\Delta H > h\nu$

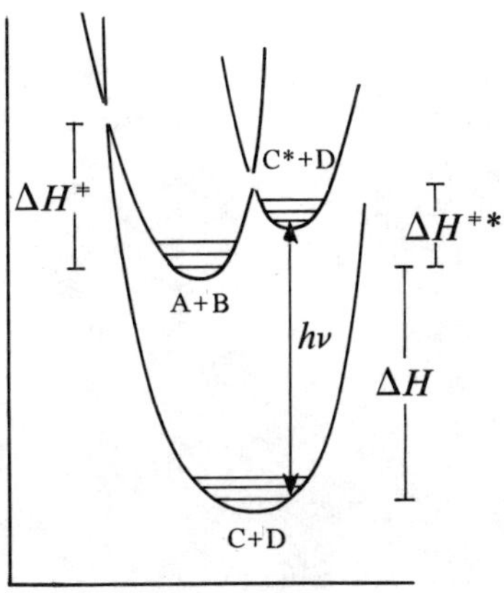


Chemiluminescent
reaction
$\Delta H < h\nu$

**Fig. 6.7** Reaction coordinate diagrams for chemical excitation processes; A + B = reactants; C + D = products formed in ground states; C* + D = products formed with C in an excited state and D in the ground state; $\Delta H$ = energy available from the reaction according to the usual thermodynamic criteria; $\Delta H^{\ddagger}$ = activation energy for formation of products in the ground state; $\Delta H^{\ddagger *}$ = activation energy for formation of one product in an excited state; $h\nu$ = energy necessary for the excitation C → C*. (Reprinted from *Accounts of Chem. Res.*, 2, 301 (1969). Copyright 1969 by the American Chemical Society. Reprinted by permission of the copyright owner.)

modes at its disposal. Considerations of this sort have lead to the design of new chemiluminescent systems, among them, the following reaction,

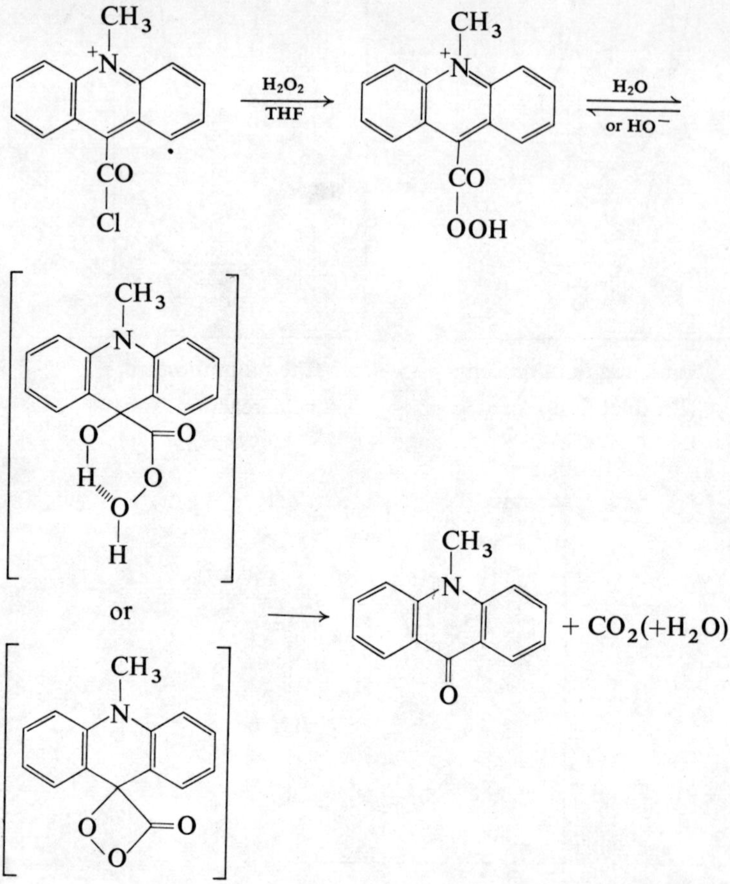


It is convenient to discuss chemiluminescent reactions under three separate headings, namely, the formation of excited oxygen, peroxide decomposition, and electron transfer.

### *6.2.3 Formation of excited oxygen*

Recently, Kasha[11] has advanced a theory of chemiluminescence involving excited states of oxygen, which nicely unites the chemiluminescence of some organic molecules in oxidation reactions and certain inorganic chemiluminescent systems. Examination of the emission spectra of the $H_2O/Cl_2$ and $H_2O_2/ClO^-$

reactions, shows the presence of bands corresponding to the following transitions:

| Transition | $\lambda_{max}$ (nm) |
| --- | --- |
| (a) $^1\Delta_g \rightarrow {}^3\Sigma_g^-$ | 1270·0 |
| (b) $^1\Sigma_g^+ \rightarrow {}^3\Sigma_g^-$ | 762·0 |
| (c) $2[^1\Delta_g] \rightarrow {}^3\Sigma_g^- + {}^3\Sigma_g^-$ | 633·5 |
| (d) $[^1\Delta_g + {}^1\Sigma_g^+] \rightarrow {}^3\Sigma_g^- + {}^3\Sigma_g^-$ | 478·0 |

Transitions (c) and (d) correspond to emission from a double molecule, generated by the interaction of two singlet oxygen molecules. Emission from the $2[^1\Sigma_g^+]$ state is also possible, but it is believed that in the $H_2O_2/Cl_2$ system this is masked by $Cl_2$ absorption.

If, however, a species is present, say an organic molecule, with favourable energy levels, direct energy transfer from the excited oxygen molecular pairs may occur, resulting in a 'sensitized' chemiluminescence. This is shown with examples in Fig. 6.8.

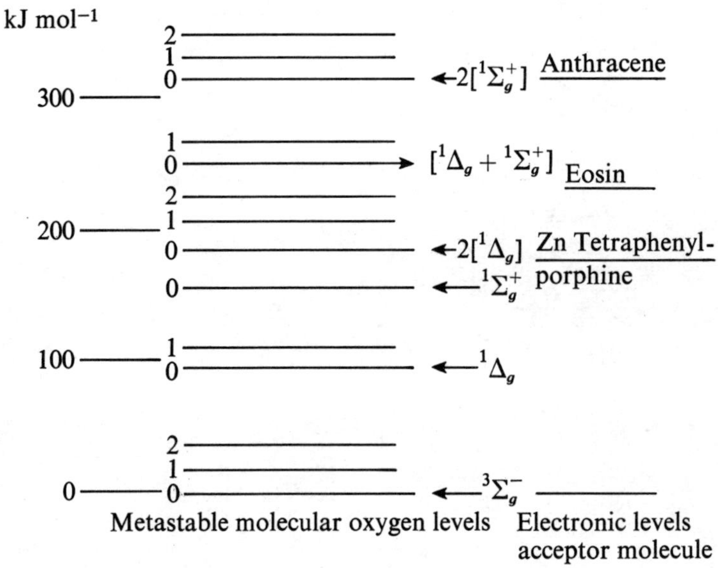


Fig. 6.8

Such a process, however, demands a squared dependence upon peroxide concentration and, while this is shown by some organic molecules, there are many that do not fulfil this requirement.

### 6.2.4 *Concerted peroxide decompositions*[10]

Many of the classical chemiluminescent reactions can be considered as examples of concerted peroxide decompositions. The essential features of the chemiluminescent reaction between 3-aminophthalhydrazide (luminol) and potassium persulphate in basic solution are summarized in Table 6.2.

TABLE 6.2: The chemiluminescent reaction between luminol, potassium persulphate, and hydrogen peroxide in basic solution

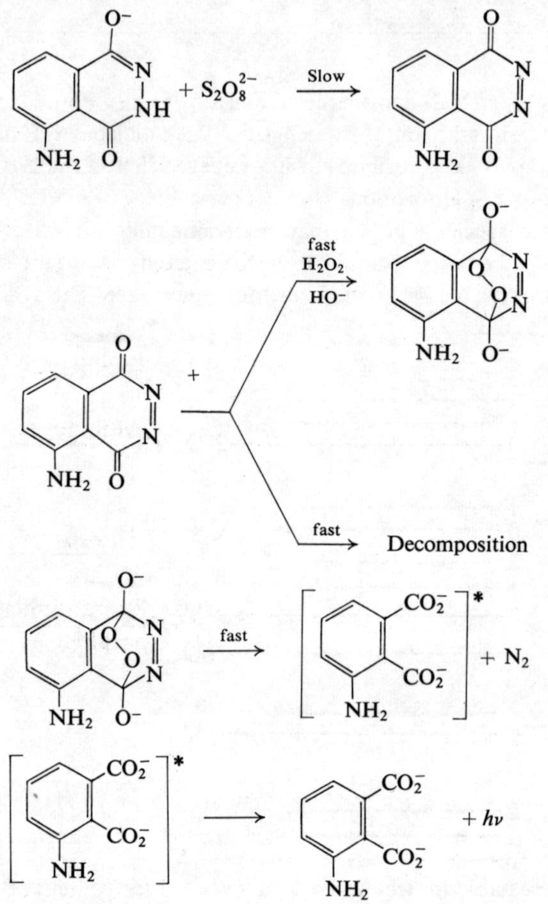


However, until recently, while it has long been recognized that decomposition of the peroxide was highly exothermic, no suggestion had been made as to why it should lead to an *electronically* excited product. The assumption that the decomposition step is concerted, leads to a recognition of the fact that such a reaction is a reverse cycloaddition involving $4\pi$ electrons and which, if orbital symmetry is preserved, requires that one of the product carbonyl groups be electronically excited.[12]

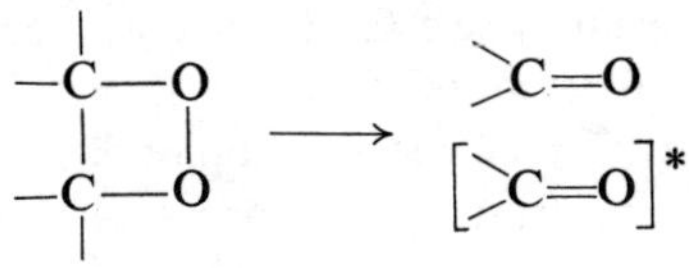

An alternative suggestion has also been made that the transition state is antiaromatic in that it contains $4\pi$ electrons and is, therefore, likely to generate an electronically excited product.

Other mechanisms of peroxide decomposition may be possible but the above proposals are attractive.

### *6.2.5 Radical-ion reactions*[13]

Photoexcitation of a molecule **A** involves transferring an electron from the highest occupied molecular orbital of **A** to the lowest unoccupied molecular orbital. A similar process can occur if the radical anion $\mathbf{A}^{\bar{\cdot}}$ is allowed to interact with the radical cation $\mathbf{A}^{\dot{+}}$ (Fig. 6.9).

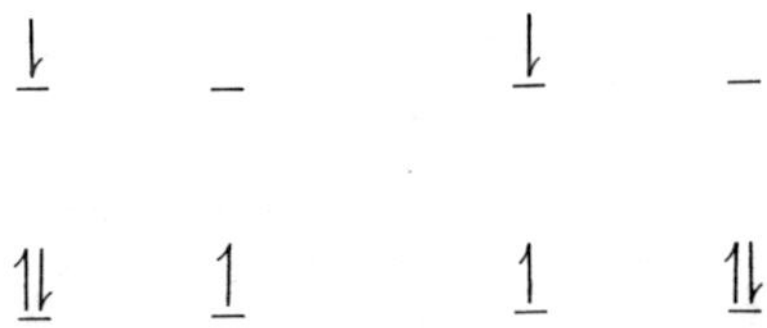

$$\mathbf{A}^{\bar{\cdot}} + \mathbf{A}^{\dot{+}} \longrightarrow {}^{1}\mathbf{A}^{*} + \mathbf{A}$$

**Fig. 6.9**

Such a reaction may be realized experimentally by electrolysing a solution of **A**, when a species $\mathbf{A}^{\bar{\cdot}}$ will be generated at the cathode and will then diffuse into solution. Reversal of the current brings about production of $\mathbf{A}^{\dot{+}}$ which, similarly, will diffuse into solution, to interact with $\mathbf{A}^{\bar{\cdot}}$ and, in favourable circumstances, will cause chemiluminescence.

## 6.3 Vision[14,15]

### *6.3.1 Structural considerations*

The human eye is a highly versatile light detector, possessing a sensitivity range of about thirteen orders of magnitude and which, at normal visual intensities, can discriminate between sources whose light intensities vary by 2%. Part of the inner lining of the eyecup embodies a thin layer of tissue known as the retina, itself comprising several layers, and among which is a layer of receptor cells. These are of two types, rods, whose main function is reception in dim light, and cones, which operate in bright light and are also responsible for colour vision. Nerve fibres run from these cells via the optic chiasma to the optic nerve and finally to the visual cortex of the brain.

The photoreceptor cells are in part made up of discs or sacs and contain within the sac membrane the purple visual pigment, rhodopsin (mol. wt. = 40 000) The chromophoric group, 11-*cis*-retinaldehyde (11-*cis*-vitamin A, aldehyde, also known as 11-*cis*-retinene) lies in the plane of the sac surface such that its orientation maximizes the absorption efficiency.

### 6.3.2 *Visual pigments*

The visual pigments are best characterized by their u.v. spectra; all show three regions of absorption as detailed below.

(a) α-Band: $\lambda_{max}$ lies between 433 and 620 nm. This decreases in intensity on irradiation.

(b) β-Band: $\lambda_{max}$ lies between 340 and 370 nm. This band increases in intensity on irradiation.

(c) γ-Band: $\lambda_{max}$ 280 nm. This is essentially a protein absorption and is insensitive to irradiation.

The visual pigments, of which four are at present known are complexes of a lipoprotein and a chromophoric group (Table 6.3).

TABLE 6.3: Structure of the rod visual pigments and chromophoric groups

| Pigment | Lipoprotein | Chromophoric group | $\lambda_{max}$ (nm) | Origin |
|---|---|---|---|---|
| Rhodopsin | Opsin | 11-*cis*-Retinaldehyde | 430–560 | Frog |
| Porphyropsins | Opsin | 3-Dehydro-11-*cis*-retinaldehyde | 520–640 | Freshwater fish |
| Iodopsin | Photopsin | 11-*cis*-Retinaldehyde | 562 | Chicken |
| Cyanopsin | Photopsin | 3-Dehydro-11-*cis*-retinaldehyde | 620 | Tadpole |

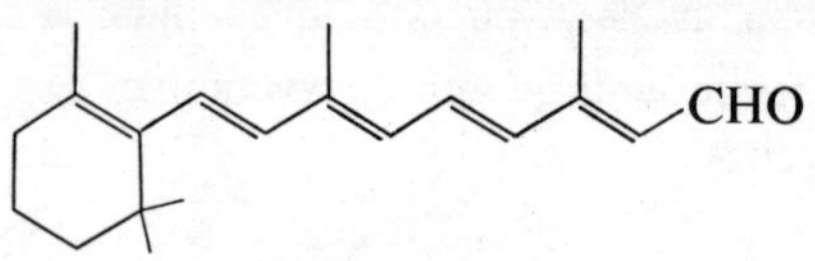

all-*trans*-Retinaldehyde
$\lambda_{max}$ = 381 nm

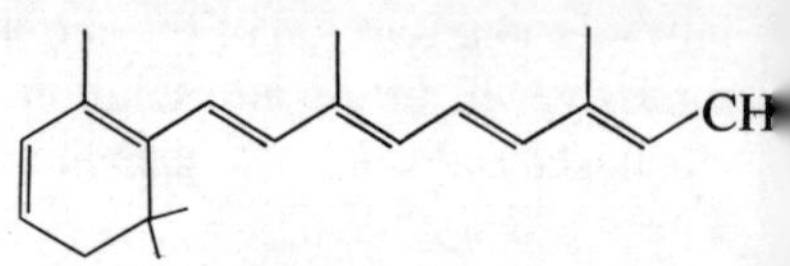

all-*trans*-3-Dehydroretinaldehyde
$\lambda_{max}$ = 401 nm

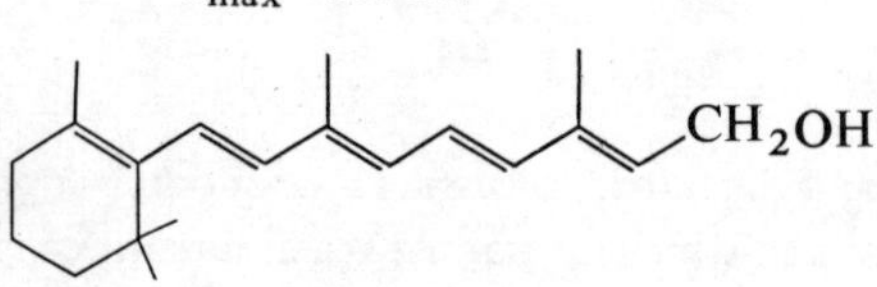

all-*trans*-Retinol
$\lambda_{max}$ = 325 nm

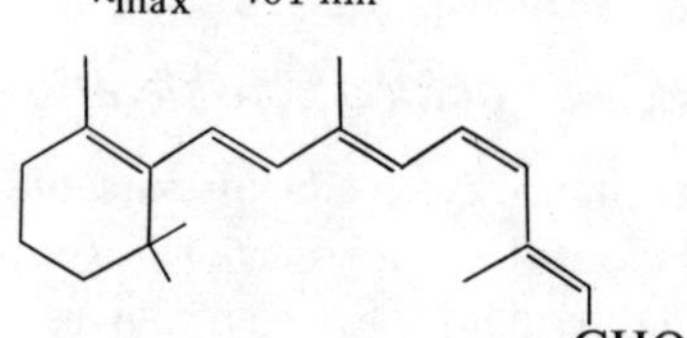

11-*cis*-Retinaldehyde
$\lambda_{max}$ = 369 nm

The way in which 11-*cis*-retinaldehyde is bound to the opsin is only partially understood. Some work[16] appears to suggest a Schiff base link, but which, if present, cannot be the only source of binding.

### 6.3.3 The photolytic process

The pathway leading to visual excitation comprises one photochemical step followed by several thermal steps. The quantum yield of the primary photochemical process has been determined by measuring the rate of bleaching of frog rhodopsin extracts at 502 nm and a figure of 0·5 arrived at.[17] It is believed that the photochemical step is the isomerization of 11-*cis*-retinaldehyde to all-

| Substance | $\lambda_{max}$ (nm) |
|---|---|
| Rhodopsin | 498 |
| $h\nu$ ↓ ↑ | |
| Prelumirhodopsin | 543 |
| ↓ above −140°C | |
| Lumirhodopsin | 497 |
| ↓ above −40°C | |
| Metarhodopsin I | 478 |
| ↓ ↑ above −15°C | |
| Metarhodopsin II | 380 |
| ↓ | |
| Metarhodopsin III | 465 |
| ↓ | |
| *N*-Retinylidene-opsin II | 440 |
| ↓ | |
| all-*trans*-Retinaldehyde + opsin | 387 |

**Fig. 6.10** Scheme for photolytic process of rhodopsin.

*trans*-retinaldehyde while bound to opsin. According to Wald,[18] this change in geometry exposes the chromophoric group to hydrolytic attack, and in this way initiates its removal from the pigment (Fig. 6.10).

The all-*trans*-retinaldehyde is finally converted to vitamin $A_1$ by an alcohol dehydrogenase and NADH.

That visual excitation requires $5 \times 10^{-2}$ s, that the generation of metarhodopsin I takes about $4 \times 10^{-3}$ s and that its decomposition occupies several minutes, strongly suggests that the conversion, metarhodopsin I to metarhodopsin II, is the process responsible for vision.

Clearly, following the photolytic step, a process of rhodopsin regeneration is necessary. Through the agency of an isomerase, vitamin $A_1$ is converted to 11-*cis*-vitamin $A_1$, which, after oxidation to 11-*cis*-retinaldehyde, rapidly combines with opsin. In this way the rhodopsin is reconstituted and the photolytic process can begin again. These main steps are shown in Fig. 6.11

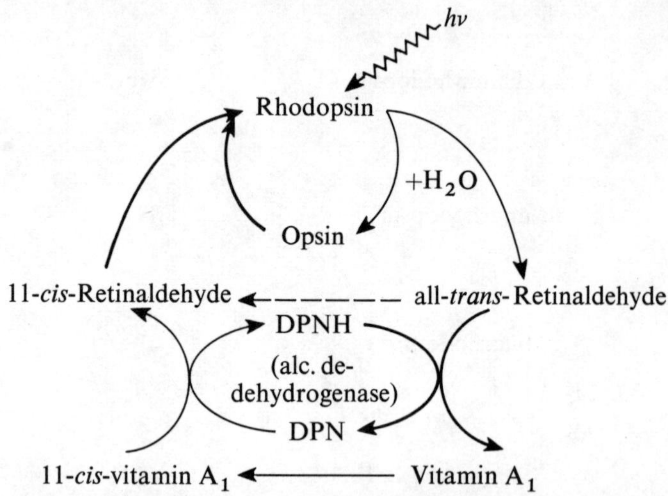


**Fig. 6.11** Cycle of light and dark reactions involving rhodopsin.

from which it will be noted that a direct transformation of all-*trans*-retinaldehyde to 11-*cis*-retinaldehyde is indicated. This is to accommodate the physiological observation that reconstitution of rhodopsin after exposure to a high intensity flash of short duration, occurs faster than dark adaptation after long exposure to light.

## *6.3.4 Visual excitation*

We must now enquire how light absorption by rhodopsin can lead to nervous excitation. Earlier it has been stated that the primary process quantum yield is about 0·5, but although this indicates a high degree of efficiency, considerable amplification is necessary before stimulation of the rod–bipolar synapse becomes effective. To account for this, three hypotheses have been advanced.

(a) Enzyme hypothesis[19]

According to this hypothesis, the rhodopsin is a proenzyme which on absorption of light and concomitant *cis-trans*-isomerization of retinaldehyde, exposes the catalytic centre. Consequently, generation of one molecule of active enzyme could lead to a turnover of perhaps $10^3$ molecules of substrate.

(b) Solid state hypothesis[20]

The outer structure of the rods is *quasi*-crystalline and the suggestion has been made that energy transfer by photoconduction might occur. It has been estimated that the quantum gain could be of the order of $10^4$ electrons per quantum.

(c) Ionic hypothesis[21]

This hypothesis depends upon recognition of the fact that the retina is part of

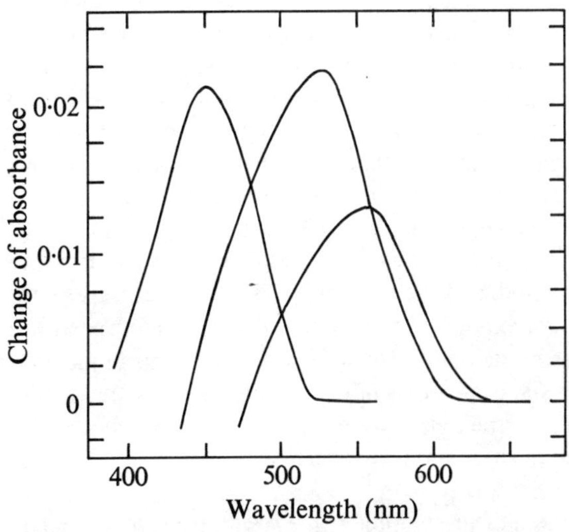


**Fig. 6.12** Difference spectra of human cone pigments.

the central nervous system, and thus, stimulation of it will lead to processes similar to those in nervous excitation. It is postulated that absorption of a photon by a rhodopsin molecule will produce, as in nervous excitation, a passive cation flux in the photoreceptor–bipolar synapse. Since photolysis of a single rhopsin molecule could lead to the exchange of many ions, the necessary amplification could be achieved.

### 6.3.5 *Colour vision*[22]

The trichromacy of colour vision, i.e., the idea that all colours can be matched by suitable mixture of three primaries is most plausibly satisfied in two ways.

(a) Each cone should contain three pigments.
(b) Each cone should contain one pigment operating in conjunction with three fixed wavelength selectors.

Resolution of this problem by analysis of the cone pigments has to date not been achieved in any mammal. However, it has been shown from experiments particularly on colour-blind subjects, that the normal eye probably contains three pigments, erythrolabe ($\lambda_{max}$ 555–565 nm), chlorolabe ($\lambda_{max}$ 550 nm) and cyanolabe ($\lambda_{max}$ 440 nm). They have the spectral distribution shown in Fig. 6.12.

## 6.4 The photographic process

Our degree of understanding of each of the separate stages in the production of a negative, that is, formation of a latent image, development and finally fixing, is opposite to that of the sequence in which they are performed. The 'reaction medium' consists of a dispersion of microcrystals of silver bromide of dimensions in the range $10^2$–$10^4$ nm in a gelatin medium which may contain in addition, a 'chemical' sensitizer such as a sulphur compound (without which the dispersion may prove inactive) and a cyanine dye to extend the range of sensitivity either to the green-yellow region (orthochromatic film) or to the red-orange region (panchromatic film). Both of these additives are adsorbed onto the crystal face, the sulphur compound forming $S^{2-}$ ions.

The blackening of a plate exposed and then developed is not related linearly to the light received, and the response is also temperature-dependent (Fig. 6.13). Except at highest exposures, the image is invisible and depends for development on the increase in the number of silver atoms at points in the latent image of between seven and nine orders of magnitude following treatment with a reducing agent. Nonetheless the sensitivity of the gelatin–AgBr emulsion is extremely high, the most sensitive grains being rendered developable following absorption of only a few quanta.

The mechanism of production of the latent image has been controversial for over thirty years and remains so. One moot point is the exact role of traces of sulphur compounds in sensitizing gelatin, the most favoured explanation invoking adsorption of $S^{2-}$ ions onto the AgBr surface to act as 'sensitivity specks' by trapping holes. One proposal around which discussion has subsequently centred is due to Gurney and Mott.[23] A quantum is absorbed by the crystal to release a mobile electron, the positive hole corresponding to a free bromine atom. The electron migrates to a defect centre, which may exist as a result either of physical dislocation or incorporation into the lattice of a foreign ion such as $S^{2-}$, and be situated either on the surface or in the body of the crystal. The second stage is neutralization of the now trapped electron by an $Ag^+$ ion which is interstitial in nature and capable of migration in the lattice to give a free silver atom. The now modified defect centre is capable of trapping a second photoelectron and subsequently a second mobile interstitial $Ag^+$ ion to give an $Ag_2$ species. Repetition will result in formation of a minute blob of silver metal consisting of the order of $10^2$ atoms on the microcrystal and visible on a photomicrograph. The holes are considered either to be much less mobile than interstitial $Ag^+$ ions, to be captured as $Br^{\bullet}$ atoms by the gelatin, to be captured at sulphide centres, or to recombine to give $Br_2$. The necessary mobility of $Ag^+$ ions in

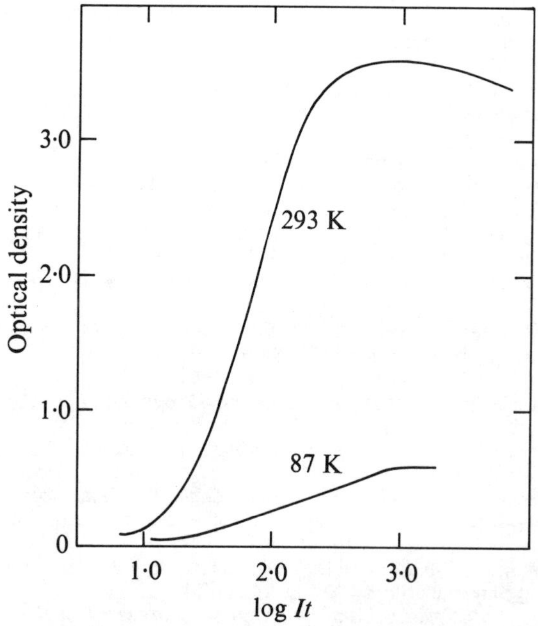


**Fig. 6.13** Density–exposure curves for pure AgBr emulsion at room temperature and 87 K.

the lattice accounts for the reduced degree of latent image formation at 87 K shown in Fig. 6.13 and is also supported by conductivity measurements, and assignment of points in the latent image to silver metal is confirmed by X-ray diffraction of heavily exposed plates displaying 'print-out'. The foregoing account is both simplified and selective and for a more detailed account of the nature of the crystal defects involved and other aspects the reader is referred to references 24 and 25.

In the development process a selective reducing agent is employed which will reduce AgBr crystals containing a latent image site and yet leave unexposed crystals unaffected. Weakly alkaline hydroquinone acts as a suitable electron source, discharging electrons into the silver blob thereby attracting and discharging further interstitial $Ag^+$ ions until the whole crystal lattice breaks up into bromide ions and metallic silver.

Fixing involves the removal of residual silver bromide by dissolution with sodium thiosulphate solution containing one of various accelerating agents.

The many technical refinements of the various stages of the photographic process are covered by John and Field.[26]

## References


1. For general accounts see (a) E. Rabinowitch and Govindjee, *Photosynthesis*, J. Wiley and Sons, Inc., New York (1969); (b) C. P. Swanson, Ed., *An Introduction to Photobiology*, Prentice-Hall, Inc., Englewood Cliffs, New Jersey (1969); (c) H. H. Seliger and W. D. McElroy, *Light: Physical and Biological Action*, Academic Press, Inc., New York (1965).
2. C. B. van Niel, *Adv. Enzymol.*, **1**, 263 (1941).
3. R. Emerson, R. V. Chalmers, and C. Cedarstrand, *Proc. Nat. Acad. Sci. U.S.*, **43**, 133 (1957).
4. M. Calvin, *Rev. Mod. Phys.*, **31**, 147 (1959).
5. R. Emerson and W. Arnold, *J. Gen. Physiol.*, **16**, 191 (1932).
6. Y. De Konchkovsky and P. Joliot, *Photochem. and Photobiol.*, **6**, 567 (1967).
7. R. S. Becker and M. Kasha, *J. Amer. Chem. Soc.*, **77**, 3669 (1955).
8. F. McCapra, *Quart. Rev. (Lond.)*, **20**, 485 (1966).
9. K. D. Gundermann, *Angew. Chem. Intern. Ed. Engl.*, **4**, 566 (1965).
10. M. M. Rauhut, *Acc. Chem. Res.*, **2**, 80 (1969).
11. A. U. Khan and M. Kasha, *J. Amer. Chem. Soc.*, **88**, 1574 (1966).
12. F. McCapra, *Chem. Comm.*, 155 (1968).
13. D. M. Hercules, *Acc. Chem. Res.*, **2**, 301 (1969).
14. R. A. Weale, in *Current Topics in Photophysiology*, Ed., A. C. Giese, Academic Press, New York, Vol. IV, p. 1 (1968).
15. S. L. Bonting, in *Current Topics in Bioenergetics*, Ed., D. R. Sanadi, Academic Press, New York, Vol. 3, p. 351 (1969).
16. R. A. Morton and G. A. J. Pitt, *Fortschr. Chem. Org. Naturstoffe*, **14**, 244 (1957).
17. C. F. Goodeve, R. J. Lythgoe, and E. E. Schneider, *Proc. Roy. Soc.*, **B130**, 380 (1942).
18. G. Wald, in *Light and Life*, Eds., W. D. McElroy and B. Glass, The Johns Hopkins University Press, Baltimore, Md., p. 742 (1961).
19. G. Wald, in *Enzymes: Units of Biological Structure and Function*, Ed., O. Gaebler, Academic Press, New York, p. 355 (1956).
20. G. Wald, P. K, Brown, and I. R. Gibbons, *J. Opt. Soc. Amer.*, **53**, 20 (1963).
21. W. A. Hagins, *Cold Spring Harbor Symp. Quant. Biol.*, **30**, 403 (1965).
22. W. A. H. Rushton, in *Current Topics on Photophysiology*, Ed., A. C. Giese, Academic Press, New York, Vol. II, p. 123 (1964).
23. R. W. Gurney and N. F. Mott, *Proc. Roy. Soc.*, **A164**, 151 (1938).
24. J. W. Mitchell and N. F. Mott, *Phil. Mag.*, Ser. 8, **2**, 1149 (1957).
25. F. C. Brown, *The Physics of Solids*, W. A. Benjamin, Chapter 12 (1967).
26. D. H. O. John and G. T. J. Field, *A Textbook of Photographic Chemistry*, Chapman and Hall, London (1963).

# Index

*Printed by William Clowes & Sons Limited, London, Colchester and Beccles*